Asymptotic Perturbation Theory of Waves

Asymptotic Perturbation Theory of Waves

Lev Ostrovsky

ICP

Imperial College Press

Published by

Imperial College Press
57 Shelton Street
Covent Garden
London WC2H 9HE

Distributed by

World Scientific Publishing Co. Pte. Ltd.
5 Toh Tuck Link, Singapore 596224
USA office: 27 Warren Street, Suite 401-402, Hackensack, NJ 07601
UK office: 57 Shelton Street, Covent Garden, London WC2H 9HE

Library of Congress Cataloging-in-Publication Data
Ostrovskii, L. A., author.
Asymptotic perturbation theory of waves / Lev Ostrovsky, NOAA ETL, USA.
pages cm
Includes bibliographical references and index.
ISBN 978-1-84816-235-8 (hardcover : alk. paper)
1. Wave-motion, Theory of. 2. Perturbation (Mathematics) 3. Nonlinear wave equations.
4. Differential equations--Asymptotic theory. I. Title.
QC157.O777 2014
531'.113301--dc23

2014023528

British Library Cataloguing-in-Publication Data
A catalogue record for this book is available from the British Library.

Printed in Singapore

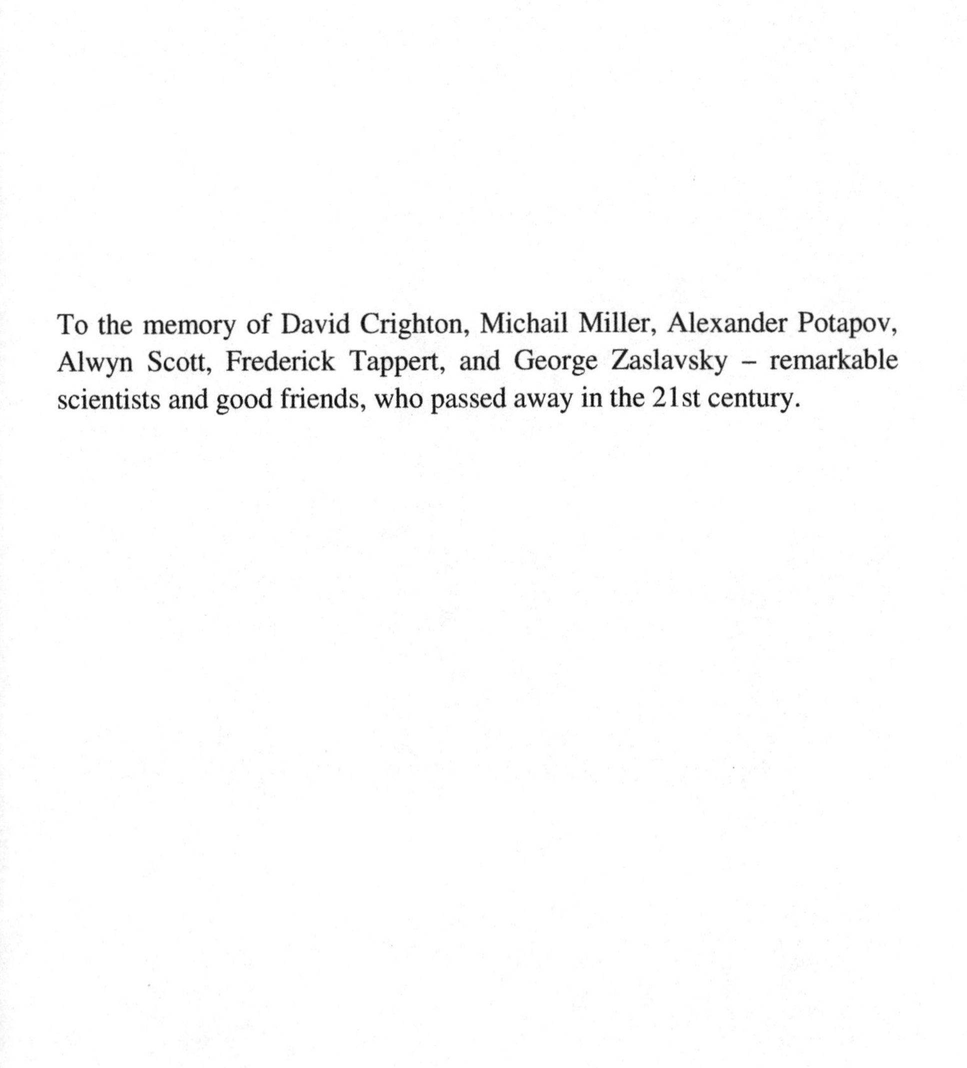

To the memory of David Crighton, Michail Miller, Alexander Potapov, Alwyn Scott, Frederick Tappert, and George Zaslavsky – remarkable scientists and good friends, who passed away in the 21st century.

Preface

> As far as the laws of mathematics refer to reality, they are not certain, and as far as they are certain, they do not refer to reality.
>
> Albert Einstein

Using approximations to find solutions of differential equations has a long history, especially in the areas of applied science and engineering, where relatively simple ways to obtain quantitative results are crucial. Since the 18th century, perturbation methods have been actively used in celestial mechanics for describing planetary motion as a three-body interaction governed by gravitation. As was shown by Poincaré [19], the three-body problem is not completely integrable, and its solution was often based on finding a small perturbation to the solution of the integrable two-body problem, when the mass of at least one of the three bodies is small. One famous example is the prediction of an existing and the position of a new planet by Adams and Leverrier based on calculation of perturbations of Uranus, which resulted in the discovery of the planet Neptune by Galle in 1846. A more recent overview of this topic can be found, for example, in [6, 7]. In the 20th century the notion of a "theory of oscillations" as a unifying concept, meaning the application of similar equations and methods of their solution to quite different physical problems, came into being. In particular, the phase plane method and Poincaré mapping proved to be very efficient. For early work in this area, see the classic book by Andronov *et al.* [3].

Regular perturbation schemes allow one to find a deviation from the basic, unperturbed solution, when the deviation remains sufficiently small indefinitely or at least at a time interval at which the solution should be determined. However, in many important cases, a small initial perturbation can strongly change the solution even if the basic equations are only slightly disturbed. In these cases the smallness of the expansion parameter is reflected in the slowness, but not necessarily smallness, of the deviation from the solution existing when that parameter is zero. The slowness means that the time of significant deviation from the unperturbed solution is much longer than the characteristic time scale of the process, such as the period of oscillations or the duration of a pulse. An adequate tool for solving such problems is the asymptotic perturbation theory, which constructs a series in a small parameter in which the main term of the expansion differs from the unperturbed solution in that it contains slowly varying parameters; their variation can be found from "compatibility," or "orthogonality," conditions which secure the finiteness of the higher-order perturbations. In the framework of an asymptotic method, the series may not even converge, but a sum of a finite number of terms in the series approaches the exact solution when the expansion parameter tends to zero, which is sufficient in most applications. Still more involved is singular perturbation theory, in which the asymptotic series does not converge to the unperturbed solution in the total interval in time or space when the small parameter tends to zero; in these cases the matched asymptotic expansions are used, for example, for the boundary layers in hydrodynamics.

As an early example of such an approach we mention the Poincaré–Lindstedt method applied to a differential equation (Duffing equation) with a small nonlinear term in which the frequency of a basic harmonic solution should be perturbed to avoid a cumulative deviation from a harmonic solution of the unperturbed linear equation [19]. In the 20th century the area of application of perturbation theories has greatly broadened, due particularly to the development of radio electronics and radiophysics. The systematic development of asymptotic perturbation schemes can be attributed to the work of Krylov and Bogoluybov [14]. An excellent description of the main ideas and applications can be found in the book by Bogolyubov and Mitropolsky [5]; for the later development see, for example, [4, 11, 15].

In spite of many variations of the asymptotic theory for ordinary differential equations (ODEs), its idea is rather general and consists of the following main steps.

1. A system of ODEs containing a small parameter $\mu << 1$ is considered. It is supposed that at $\mu = 0$ this system has a family of periodic solutions, $u = U(\theta, A)$, where $\theta = \omega t$ and ω and A are constants; in particular, but not necessarily, the function U can be a sinusoid so that ω is its frequency and A is its amplitude.
2. At $\mu \neq 0$ the solution is expanded in a power series of μ, and the parameters ω and A depend on the "slow" parameter $T = \mu t$. This series is substituted into the equations to be solved.
3. These equations are, in turn, expanded in a series of μ. As a result, in each order of μ we obtain linear, inhomogeneous equations, with a right-hand side ("forcing") that is known from the solution obtained in previous approximations.
4. The solution of these equations is, in general, secularly (i.e., as a power of time) divergent, and to keep the perturbations finite, additional "orthogonality" conditions should be imposed, which are typically differential equations determining a slow time variation of the parameters, including the functions $A(T)$ and $\omega(T)$. This defines first the variation of U and then the higher-order perturbations.

This procedure can be complicated; for example, the solution can depend on more than one phase θ_i and there can be several parameters A, but the concept of the method remains essentially the same. Note that due to the use of "two times," t and T, such a scheme is also called the method of multiple scales.

Various approximate approaches for *waves* described by partial differential equations (PDEs) have been used since very early as well; it suffices to mention the use of the ray concept in optics. Intensive use of perturbation methods in the theory of waves began in the 20th century, in relation to quantum mechanics methods such as the "quasi-classical" WKB theory used in linear boundary problems [4] and, for nonlinear waves, to physical oceanography, nonlinear optics, and many other areas.

The general "theory of waves" understood in the same sense as "theory of oscillations" was formulated in the mid-1900s; see [12, 17, 20, 21].

Since the 1960s, the success of the mathematical theory of waves was largely associated with exact methods such as the inverse scattering method in the theory of integrable nonlinear wave equations [2, 16]. Another prominent result obtained, in part, due to the advancement of computers, was the discovery of the particle-like behavior of interacting solitary waves by Zabusky and Kruskal, which motivated them to name them solitons [22]. These achievements allowed the development of a new chapter of the PDE theory.

However, the class of exactly integrable, nonlinear PDEs is, as a rule, the result of an approximation (more specifically, reduction) with respect to a basic physical system, obtained by using the smallness of such factors as nonlinearity and dispersion. The well-known example is the Korteweg–de Vries (KdV) equation suggested in 1895 for the approximate description of shallow-water waves propagating in a given direction. There exist many other "model" or "evolution" equations, typically obtained as an approximation of more general physical equations. By now, numerous books and review papers are available which describe these results, especially with regard to the inverse scattering theory; see, for example [2, 9, 16].

From a practical viewpoint, completely integrable equations are often too strong an idealization in which such important factors as dissipation, inhomogeneity, and others, are ignored. Moreover, even when the exact methods work, the resulting solution can be too cumbersome for practical use, and the corresponding numerical calculations may be almost as time consuming as the numerical solution of the basic equations, with no simple physical interpretation. Thus, perturbation methods remain important even in these cases. Some of these methods use the closeness of the equations to integrable ones (see [13] and references therein). We shall briefly outline such methods in Chapter 6. They can be as complicated as the inverse scattering method itself, and were developed only for solitons. Here, we are concerned mainly with the "direct" methods of a solution of PDEs based on the same principles as those known for ODEs in the oscillation theory, as outlined above. For a PDE containing a small parameter (which can now be present not only in the equations but in the initial or boundary conditions as well), it is supposed that at $\mu = 0$ there exists a family of exact solutions, for example, periodic solutions such as $u = U(\theta, A)$, where now $\theta = \omega t - kx$. Then at $\mu \neq 0$ the solution can be constructed by using a method consisting of steps which are essentially similar to the steps 1–4 described above for the ODEs. Another class of solutions involves localized pulses such as solitons and fronts (kinks, shock waves, excitation fronts), at $\mu = 0$ having the form $u = U(\zeta, A)$, where $\zeta = x - Vt$ with a constant velocity V, and U is a localized or transient function.

Hence, this approach covers a variety of processes, from oscillations to periodic waves to solitary waves to shocks to autowaves. In spite of such a variety, the principle of obtaining the approximate solution is essentially the same.

A direct approach in the theory of nonlinear waves has been developed since the early 1970s; see the history and references in [18]. Many relevant materials were included in other books and review papers, mainly regarding solitons [1, 2, 8]. In the book [18] the concept of "modulated waves" with slowly varying parameters, both quasi-periodic and solitary, along with other phenomena, history, and problems for solving, was described using mainly a semi-phenomenological approach. There are apparently no books that systematically discuss the asymptotic perturbation schemes for various classes of waves, in both integrable and non-integrable equations; as will be shown, the latter difference is not crucial for the direct method.

This book describes asymptotic perturbation schemes for periodic and solitary waves, both in their general form and as they apply to particular equations. The book is organized as follows.

The first four chapters deal with quasi-periodic wave processes. Chapter 1 is introductory; it contains illustrative examples of perturbed ODEs and PDEs and the use of regular and asymptotic perturbation approaches to their solution. In Chapter 2 we first present a perturbation scheme for quasi-harmonic waves in a general system of weakly nonlinear equations and then consider some examples of its application to linear and nonlinear waves. In Chapter 3 this approach is generalized to non-harmonic periodic waves in possibly strongly nonlinear systems, and then applied to Lagrangian systems which allows, in particular, the derivation of the Whitham's averaged variational principle. In Chapter 4 the general results are applied to specific nonlinear wave processes, including the evolution of a stepwise initial condition into a train of nonlinear oscillations.

The following four chapters deal with impulses and fronts. Chapter 5 presents a general perturbation scheme for localized nonlinear waves: solitons and fronts. In Chapter 6 this method is applied to studying the evolution of solitons perturbed by dissipation, amplification, and refraction. In Chapter 7 we consider various classes of interaction between solitons and between kinks. The final chapter concentrates mainly on active processes such as parametric amplification of solitons by a long wave and formation of "autowaves" in reaction-diffusion systems. Another distinctive feature of Chapters 7 and 8 is that they describe interaction of waves of different spatial scales by matching the "fast" and "slow" motions.

In most examples considered here, the equations to be solved are taken without derivation, with only a brief reference to their physical origin. Detailed derivation of similar equations in various physical and biological problems can be found in many books, such as [8, 12, 18, 20, 21]. Note also that this book does not pursue mathematical strictness; there are no theorems and lemmas in the book. Our main goal is to explain how the corresponding asymptotic perturbation methods are constructed, how they can be applied to specific models, and what the qualitative features of the resulting approximate solutions are. Many other relevant applications can be found in [18]. Also, our reference lists in the chapters are limited; they cite books and articles used in the material of the chapter, and some relevant references, including historical ones. In the books cited at the end of the Preface, the reader can find many other relevant references.

The author hopes that this small book will convince the reader that the direct asymptotic perturbation method is efficient in treating a variety of equations describing linear and nonlinear waves, the parameters of which vary slowly, at sufficiently long scales in time and/or space. In some cases the asymptotic scheme can be constructed in a mathematically stricter way, the convergence can be proven, etc.;

however, our aim is to demonstrate the universality of such schemes and their effectiveness in various, often seemingly unconnected, applications.

Since the book contains both tutorial and research elements, it may be of interest to a broad audience, from graduate students to professionals in applied mathematics and different areas of physics.

The author is grateful to K. A. Gorshkov and Yu. A. Stepanyants for the extremely helpful discussions regarding both the content and structure of the book, and to B. A. Malomed for the helpful notes concerning the book plan. In addition to the published sources, in parts of Chapters 6 and 7, some materials from the doctoral thesis of K. A. Gorshkov [10] were used with his kind consent. Many thanks to S. Westgaard, who helped to improve the style of the manuscript, and R. Babu, M. Judge and T. D'Cruz for their crucial technical help. Note also that the idea of this book was discussed with G. M. Zaslavsky who more than once stressed the importance of its topic.

Lev Ostrovsky
Boulder, Colorado, 2014

References

1. Ablowitz, M. J. (2011). *Nonlinear Dispersive Waves: Asymptotic Analysis and Solitons.* Cambridge University Press, Cambridge.
2. Ablowitz, M. J. and Segur, H. (1981). *Solitons and the Inverse Scattering Transform.* SIAM, Philadelphia.
3. Andronov, A. A., Vitt, A. A. and Khaikin, S. E. (1966). *Theory of Oscillators.* Pergamon, Oxford.
4. Bender C. M. and Orsag, S. A. (1978). *Advanced Mathematical Methods for Scientists and Engineers: Asymptotic Methods and Perturbation Theory.* McGraw Hill, New York.
5. Bogolyubov, N. N. and Mitropolsky, Yu. A. (1961). *Asymptotic Methods in the Theory of Non-Linear Oscillations.* Gordon and Breach, New York.
6. Brouwer, D., G. and Clemence, G. (1961). *Methods of Celestial Mechanics.* Academic Press, New York.
7. Celletti, A. (2010). *Stability and Chaos in Celestial Mechanics.* Springer-Praxis, XVI.

8. Debnath, L. (2005). *Nonlinear Partial Differential Equations for Scientists and Engineers*. 2nd Ed. Birkhauser, Boston.
9. Dodd, R. K., Eilbeck, J. C., Gibbon, J. D. and Morris, H. C. (1982). *Solitons and Nonlinear Wave Equations*. Academic Press, London, New York.
10. Gorshkov, K. A. (2007). *Perturbation theory in soliton dynamics*. Doctor of Sci. thesis, Institute of Applied Physics, Nizhny Novgorod, 2007 (in Russian).
11. Hinch, E. J. (1991). *Perturbation Methods*. Cambridge University Press, Cambridge.
12. Karpman, V. I. (1975). *Nonlinear Waves in Dispersive Media*. Pergamon, Oxford.
13. Kivshar, Yu. S. and Malomed, B. A. (1989). Dynamics of solitons in nearly integrable systems. *Rev. Modern Physics*, v. 61, pp. 763–915.
14. Krylov, N. M., and Bogolyubov, N. N. (1947) (in English, partial translation from Russian). *Introduction to non-linear mechanics*. Princeton University Press, Princeton.
15. Nayfeh, A. H. (1973). *Perturbation Methods*. John Wiley, New York.
16. Novikov, S. P., Manakov, S. V., Pitaevskii, L. P., and Zakharov, V. E. (1976). *Theory of Solitons: The Inverse Scattering Method*. Plenum Press, New York, London.
17. Ostrovsky, L. A. and Gorshkov, K. A. (2000). Perturbation theories for nonlinear waves, in *Nonlinear Science at the Dawn at the XXI Century*, eds. P. Christiansen, M. Soerensen, and A. Scott. Springer, Berlin.
18. Ostrovsky, L. A. and Potapov, A. I. (1999). *Modulated Waves: Theory and Applications*. Johns Hopkins University Press, Baltimore.
19. Poincare, H. (1892). *Les Méthodes Nouvelles de la Méchanique Céleste*. Gauthier Villars, Paris.
20. Rabinovich, M. I., and Trubetskov, D. I. (1989). *Oscillations and Waves in Linear and Nonlinear Systems*. Kluwer, Dordrecht.
21. Whitham, G. B. (1974). *Linear and Nonlinear Waves*. John Wiley, New York.
22. Zabusky, N. J. and Kruskal, M. D. (1965). Interaction of solitons in a collisionless plasma and the recurrence of initial states. *Phys. Rev. Lett.* v. 15, pp. 240–243.

Contents

Chapter 1

Perturbed Oscillations and Waves: Introductory Examples

> Mathematics in general is fundamentally the science of self-evident things.
>
> Felix Klein

In this chapter, the main ideas will be explained by using simple examples. These examples refer to both ODEs and PDEs; in other words, to oscillations and waves. They illustrate a difference between "regular" and "asymptotic" perturbation methods as described in the Preface. In particular, the important role of resonance is demonstrated, due to which small terms in the governing equation can result in growing deviations of the solution from the unperturbed solution. Fortunately, the resonant perturbations typically have the same structure as the basic solution, and they can be incorporated into the latter by allowing its parameters (amplitude, period) to slowly vary in time, thus keeping the remaining non-resonant perturbations small. This chapter deals with the examples of quasi-harmonic solutions to weakly nonlinear equations; more complex processes will be considered further in the book.

1.1 Quasi-Harmonic Oscillators

1.1.1. *Linear oscillator with damping*

Secular growth of perturbations can be observed even in a simple linear equation of a dissipative oscillator:

$$\ddot{x} + \omega_0^2 x = -\mu g \dot{x}, \tag{1.1}$$

where ω_0 and g are constants and the small parameter $\mu \ll 1$ characterizes smallness of dissipation. To be specific, we choose the initial conditions as follows:

$$x(0) = A_0, \quad \dot{x}(0) = 0. \tag{1.2}$$

In this case the exact solution can easily be obtained:

$$x = A_0 e^{-\mu g t/2} \left[\cos\left(t\sqrt{\omega_0^2 - g^2\mu^2/4}\right) + \frac{\mu g}{2} \frac{\sin\left(t\sqrt{\omega_0^2 - g^2\mu^2/4}\right)}{\sqrt{\omega_0^2 - g^2\mu^2/4}} \right]. \tag{1.3}$$

Along with this solution, let us obtain an approximate one, supposing that the damping term with μ is small as compared with each term in the left-hand side. In this case the solution can be sought in the form of a series:

$$x(t) \approx x_0(t) + \mu x_1(t) + \mu^2 x_2(t) + \dots. \tag{1.4}$$

In the zero-order (main) approximation, letting $\mu = 0$ in (1.1) or (1.3), we have a harmonic solution:

$$x_0(t) = A_0 \cos \omega_0 t. \tag{1.5}$$

At $\mu \neq 0$, substitution of the series (1.4) into (1.1) yields in the first order of μ:

$$\ddot{x}_1 + \omega_0^2 x_1 = g\omega_0 A_0 \sin \omega_0 t \tag{1.6}$$

with the solution

$$x_1 = -\frac{gA_0}{2\omega_0}\left(\sin \omega_0 t - \omega_0 t \cos \omega_0 t\right). \tag{1.7}$$

Hence, an approximate solution combining the zero- and first-order terms is

$$x = A_0\left(1+\frac{\mu g t}{2}\right)\cos\omega_0 t - \frac{\mu g A_0}{2\omega_0}\sin\omega_0 t. \tag{1.8}$$

This solution remains close to the main one, (1.5), only at a time interval $t << 1/g\mu$. Eventually, the secularly (in this case, linearly) growing perturbation drives the solution away from its non-perturbed version (1.5). The reason for that is *resonance*: the right-hand side of (1.6) (a "forcing") has the same frequency ω_0 as the "free" solution of the homogeneous equation (1.6) at $\mu = 0$.

To avoid such a divergence, we seek a solution in the form of *modulated* oscillations:

$$x = A(T)\cos\omega_0 t + \mu x_1(t) + O(\mu^2), \tag{1.9}$$

with a slow temporal dependence of wave amplitude; here $T = \mu t$ is "slow" time, so that the first derivative $dA/dt = \mu dA/dT$ is of the order of μ, the second derivative of A is of the order of μ^2, etc. Substituting the solution (1.9) into the basic equation (1.1), we obtain in the first approximation, instead of (1.6),

$$\ddot{x}_1 + \omega_0^2 x_1 = (g\omega_0 A + 2\omega_0 A_T)\sin\omega_0 t. \tag{1.10}$$

Here and further the subscript T means differentiation with respect to slow time. The solution of (1.10) has the form

$$x_1 = \frac{1}{2\omega_0^2}(g\omega_0 A + 2A_T)(\sin\omega_0 t - \omega_0 t\cos\omega_0 t). \tag{1.11}$$

This expression still contains a secularly increasing term but now it can be eliminated by letting $2A_T + g\omega_0 A = 0$, or $A = A_0\exp(-g\omega_0 T/2)$. Thus,

$$x = A_0 e^{-gT/2}\cos\omega_0 t + O(\mu^2). \tag{1.12}$$

Now we have a limited, damping solution in the first-order approximation. Note that in this case we have an additional gain: the solution (1.12) remains valid not only in the main but also in the first approximation, because the next term is of the order of μ^2. Only in the second approximation does a perturbation appear that contains a small frequency shift of the order of μ^2. We shall add frequency perturbations below in another example.

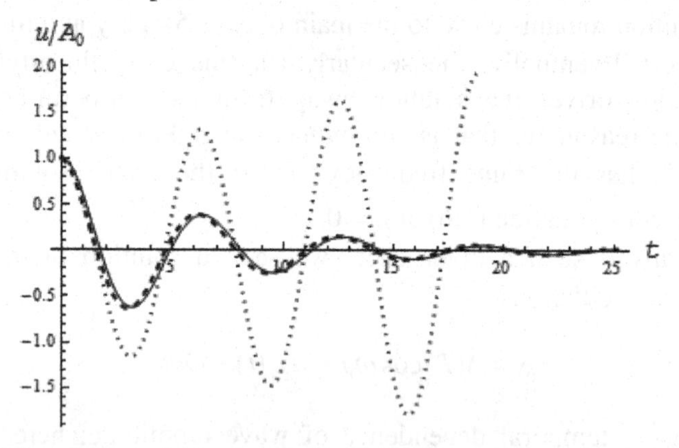

Figure 1.1. Thin solid line: the exact solution (1.3) with $\omega_0 = 1$, $g = 1$, $\mu = 0.3$. Dashed line: the asymptotic solution (1.12) with the same parameters. Gray dotted line: the perturbed solution (1.7) at $\omega_0 = 1$, $g = 1$, $\mu = 0.1$.

Figure 1.1 shows a comparison between the exact solution (1.3), the asymptotic solution (1.12), and the regularly perturbed solution (1.7). It illustrates good applicability of the asymptotic solution even for a rather strong perturbation, $\mu = 0.3$ (the very small but noticeable divergence of two curves is due to the small difference in frequencies which, as mentioned, appear in the second approximation; for smaller μ the curves are indistinguishable), and the poor behavior of the plainly perturbed solution which increases much quicker even for a smaller perturbation ($\mu = 0.1$).

1.1.2. *Oscillator with cubic nonlinearity*

We now consider a nonlinear equation often referred to as the Duffing equation:

$$\ddot{x} + \omega_0^2 x = \mu q x^3, \tag{1.13}$$

where ω_0 and q are constants and $\mu << 1$, with the same initial conditions as above:

$$x(0) = A_0, \quad \dot{x}(0) = 0. \tag{1.14}$$

The general solution of this equation can be expressed in elliptic functions [6]; here we shall analyze it at small nonlinearity ($\mu \ll 1$).

The zero-order solution x_0 is given by (1.5) as above, and after substitution of the expansion (1.4) into (1.13), the equation for the first-order perturbation $x_1(t)$ now reads

$$\ddot{x}_1 + \omega_0^2 x_1 = q x_0^3 = q A_0^3 \cos^3 \omega_0 t = q \frac{A_0^3}{4} (3\cos \omega_0 t + \cos 3\omega_0 t). \tag{1.15}$$

As a result, the approximate solution valid up to the first-order approximation is

$$x = A_0 \cos \omega_0 t + \frac{\mu q A_0^3}{32\omega_0^2} (12\omega_0 t \sin \omega_0 t + \cos \omega_0 t - \cos 3\omega_0 t). \tag{1.16}$$

Hence, small nonlinearity produces a perturbation at the basic frequency and at the third harmonic of the linear solution. As in the above case, there is the secularly increasing resonant term, proportional to $t \sin\omega_0 t$. To eliminate secular growth, it is sufficient to represent the approximate solution in the same form as given by (1.4) but with the perturbed frequency:

$$x_0(t) = A_0 \cos \omega t, \; \omega = \omega_0 + \mu \omega_1. \tag{1.17}$$

Substituting again the expansion (1.4) into (1.13), now with x_0 in the form of (1.17), we obtain in the same approximation as in (1.16),

$$\begin{aligned} \ddot{x}_1 + \omega^2 x_1 &= \left(2\omega_0 \omega_1 x_0 + q x_0^3\right) \\ &= A_0 \cos \omega t \left(2\omega_0 \omega_1 + \frac{3\mu q A_0^2}{4}\right) + \frac{q A_0^3}{4} \cos 3\omega t. \end{aligned} \tag{1.18}$$

After choosing $\omega_1 = -3qA_0^2/8\omega_0$ the term proportional to $\cos\omega t$ disappears from the right-hand side of (1.18), thus eliminating the resonance and the corresponding secular growth of the perturbation. Indeed, the solution of (1.18) is now

$$x = A_0 \cos \omega t + \frac{\mu q A_0^3}{32\omega_0^2}(\cos \omega t - \cos 3\omega t), \quad \omega = \omega_0 - \frac{3\mu q A_0^2}{8\omega_0}. \quad (1.19)$$

In this approximation the perturbation in (1.19) remains finite at all times.

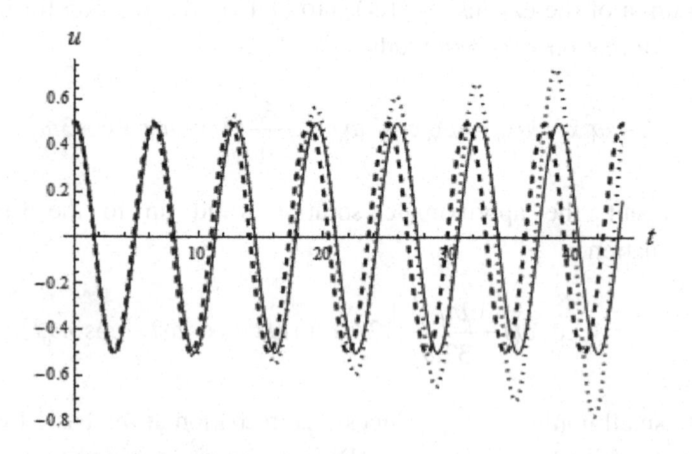

Figure 1.2. Solid line: exact solution of the Duffing equation (1.13) and the asymptotic solution (1.19) (indistinguishable). Thick dashed line: zero-order solution (1.5). Dotted line: secularly growing solution (1.16). All solutions are plotted for $\omega_0 = 1$, $q = 1$, $\mu = 0.3$, at the same initial conditions (1.14) with $A_0 = 0.5$.

Figure 1.2 shows a comparison of the exact solution of (1.13) (not given here in the explicit form) with a zero-order sinusoidal solution (1.14), a secular solution (1.15), and a non-secular asymptotic solution (1.19). The plot for the latter solution is indistinguishable from that of the exact solution.

Note that the Duffing equation with strong nonlinearity and a harmonic forcing was an object of numerous studies; in particular, it describes chaotic oscillations [5].

1.1.3. *An active system: Van der Pol oscillator*

There exists a great variety of nonlinear ODEs which are important in physics, biophysics, and engineering, and can be effectively solved with the asymptotic perturbation theory. In addition to the two above-cited examples, we briefly consider an example of an *active* system in which oscillations can be amplified and supported by some source of energy. This example is an oscillator introduced by Van der Pol [10] and then used as a model of many practical systems [1, 4]:

$$\ddot{x} + \omega_0^2 x = \mu m(1 - qx^2)\dot{x}.. \tag{1.20}$$

In this example x is assumed dimensionless, so that m has a dimension of frequency.

Using the same approach as above, we represent the solution in the form

$$x = A(T)\cos \omega t + \mu x_1 + ..., \quad \omega = \omega_0 + \mu\omega_1 + ..., \tag{1.21}$$

to obtain, after substitution into (1.20) and neglecting terms of the order of μ^2 and higher:

$$\begin{aligned} \ddot{x}_1 + \omega^2 x_1 &= 2\omega_0 \frac{dA}{dT} \sin \omega t - 2A\omega_0\omega_1 \cos \omega t \\ &-Am\omega_0 \sin \omega t + mq\omega_0 \frac{A^3}{4}(\sin \omega t + \sin 3\omega t). \end{aligned} \tag{1.22}$$

We already know that to suppress secular growth of x_1, it is necessary to eliminate resonance terms having the frequency ω. In other words, one should eliminate the terms proportional to $\sin \omega t$ and $\cos \omega t$ in the right-hand part of (1.22) by allowing

$$\frac{dA}{dT} = m\left(\frac{A}{2} - q\frac{A^3}{8}\right), \quad \omega_1 = 0, \tag{1.23}$$

which results in the following variation of the amplitude:

$$A(T) = \frac{A_0 e^{mT/2}}{\sqrt{1 + \left(qA_0^2 / 4\right)(e^{mT} - 1)}}. \tag{1.24}$$

The approximate solution is then

$$x = A(T)\cos\omega_0 t - \mu\frac{mqA^3(T)}{32\omega_0}\sin 3\omega_0 t. \tag{1.25}$$

This solution satisfies the same initial conditions (1.2) as above. From (1.24) we see that the amplitude of the main harmonic first increases and then saturates (Fig. 1.3).

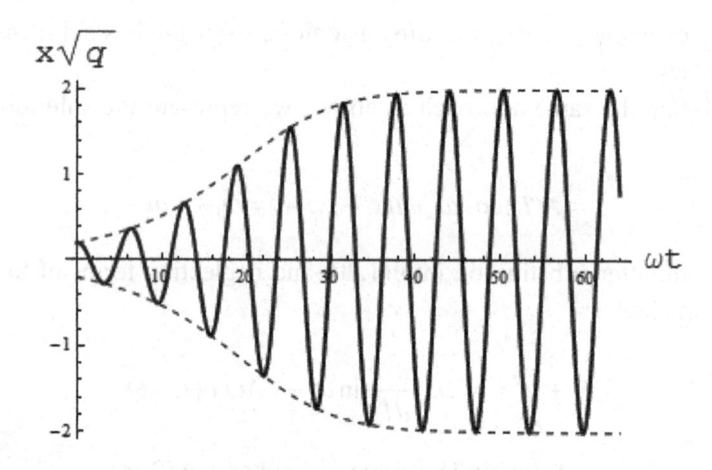

Figure 1.3. Variation of the main harmonic of the solution according to (1.24) and (1.25), at $m = 0.2$.

As is easily seen from (1.24), at $t \to \infty$ the amplitude tends to a fixed limiting value which is independent of the initial value A_0. This is a characteristic feature of an *attractor* (in this specific case, a *limit cycle*), i.e., the motion "attracting" a range of initial states of the system. This is clearly seen on the *phase plane* on which the dependence of dx/dt on x is plotted from the first integral of (1.20). This plot presents a "phase portrait" of an autonomous, second-order ODE as an image of the system motions (phase trajectories) from different initial points on the phase plane (i.e., different initial conditions). Figure 1.4 shows the phase plane of Eq. (1.20).

Among numerous applications, the Van der Pol oscillator played an important role as a realistic model of electronic generators. Moreover, the early models of a laser were based on a similar type of ODEs. Many more examples of application of asymptotic theory to ODEs can be found in [3, 4].

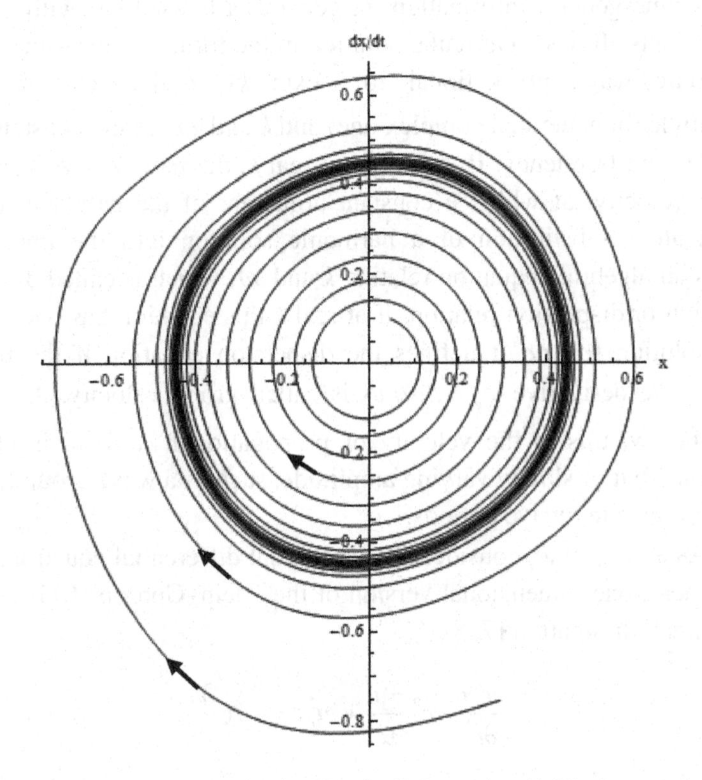

Figure 1.4. Typical phase trajectories for Van der Pol equation (1.20). Arrows show the direction of motion in time. Thick closed trajectory is the limit cycle.

1.2. Quasi-Harmonic Waves

As mentioned, similar principles of construction of asymptotic solutions are applicable for the waves, albeit with many more varieties and typically more complex mathematics, since instead of ODEs, one needs to deal with the partial-difference equations (PDEs). In this chapter we consider a few introductory examples.

1.2.1. *Linear wave equation with dissipation*

First, some general information. In general, a linear PDE with constant coefficients allows a particular solution in the form of a one-dimensional harmonic wave proportional to $a\exp[i(kx-\omega t)]$ where a is the amplitude (in general, a complex one) and k and ω are the constant wave number and frequency. If ω is real at real k, the ratio $V = \omega/k$ is called phase velocity, at which a constant argument of the exponent (phase) propagates. Substitution of a harmonic solution into the linear PDE yields an algebraic equation relating k and ω, which is called dispersion equation or dispersion relation. If at real k this equation has one or more real solution for ω, it defines the dispersion equation if the form of $\omega(k)$. The derivative $c_g = d\omega/dk$ is called group velocity. As will be seen below, this is the velocity of propagation if a wave is close to harmonic but of slowly varying amplitude, such as a wave group (packet) containing many wave periods.

As an example, consider a linear partial differential equation, which is a linear, one-dimensional version of the Klein–Gordon (KG) equation with small dissipation [7, 8]:

$$\frac{\partial^2 u}{\partial t^2} - c_0^2 \frac{\partial^2 u}{\partial x^2} + \omega_0^2 u = -\mu g \frac{\partial u}{\partial t}. \tag{1.26}$$

Unlike that which was done above for ODEs, here and further we denote the dependent (field) variable u, whereas x is now a spatial coordinate. At $\mu = 0$ this equation has a particular solution in the form of a stationary traveling (progressive) wave in which $u = u\,(x - V\,t)$ with V = const. Stationary waves are a typical basic solution on which asymptotic solutions are then constructed. Denoting $\zeta = x - Vt$, from (1.26) we have the ODE

$$\frac{d^2 u}{d\zeta^2} + \frac{\omega_0^2 u}{V^2 - c_0^2} = 0. \tag{1.27}$$

It has a harmonic solution with an arbitrary wave number k:

$$u = A\cos k\zeta,\ k^2 = \omega_0^2 / \left(V^2 - c_0^2\right). \tag{1.28}$$

It is seen that a wave propagating with a real frequency $\omega = kV$ must be "fast"; i.e., it exists if $V^2 > c_0^2$. This situation is typical of, for example, electromagnetic waves in isotropic plasma, as well as of waves propagating in an electromagnetic waveguide [7].

As mentioned, $V = \omega/k$ is the phase velocity. Substituting this into the second relation (1.28), we obtain the relationship between the wave frequency and its wave number, or the dispersion equation:

$$\omega^2 = \omega_0^2 + c_0^2 k^2. \tag{1.29}$$

The dependencies of frequency, phase velocity, and group velocity on the wave number for a wave propagating in a positive direction ($\omega > 0$ for $k > 0$) are shown in Fig. 1.5. It is seen that in a propagating wave, the frequency must exceed a definite "cutoff" value; in this case it is 1. The wave group velocity is $c_g = c_0^2 / V$, so that $c_g V = c_0^2 = \text{const}(k)$. In this case phase velocity always exceeds group velocity; they tend to each other asymptotically, at $k \to \infty$.

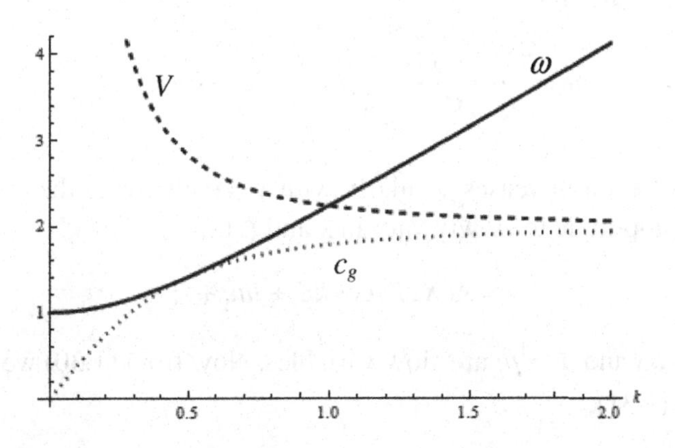

Figure 1.5. Dispersion curves according to (1.29) at $\omega_0 = 1$, $c_0 = 2$. Solid line: frequency ω. Dashed line: phase velocity V. Dotted line: group velocity c_g.

For the full Eq. (1.26) which includes dissipation, the regular perturbation scheme similar to (1.4) begins with a series:

$$u = u_0(\zeta) + \mu u_1(x,t) + \dots .$$

Substituting this into (1.26), we have

$$\frac{\partial^2 u_1}{\partial t^2} - c_0^2 \frac{\partial^2 u_1}{\partial x^2} + \omega_0^2 u_1 = g \frac{\partial u_0}{\partial t} = -g\omega A \sin k\zeta. \tag{1.30}$$

The general solution of Eq. (1.30) can be presented as the sum of a free solution satisfying (1.30) at $g = 0$ and a "forced" solution proportional to g which is a function of ζ. From the above it is clear that only the latter may present a danger of resonance. Thus, looking for the perturbation in the form of $u_1 = u_1(\zeta)$, we reduce (1.30) to an ODE:

$$\frac{d^2 u_1}{d\zeta^2} + k^2 u = -\frac{gA\omega}{V^2 - c_0^2} \sin k\zeta, \tag{1.31}$$

having the solution

$$u_1 = \frac{g\omega A}{2k^2\left(V^2 - c_0^2\right)}\left(kx \cos k\zeta - \sin k\zeta\right). \tag{1.32}$$

This perturbation increases secularly with x. To eliminate the growth we allow the amplitude to slowly vary in x and t; i.e.,

$$u = A(X,T)\cos k\zeta + \mu u_1 + \dots \tag{1.33}$$

Here $X = \mu x$ and $T = \mu t$ are slow variables. Now from (1.30) we obtain, instead of (1.31),

$$\frac{d^2 u_1}{d\zeta^2} + k^2 u_1 = \frac{-1}{\left(V^2 - c_0^2\right)}\left(g\omega A + 2kc_0^2 \frac{\partial A}{\partial X} + 2kV \frac{\partial A}{\partial T}\right) \sin k\zeta. \tag{1.34}$$

Integrating this equation with respect to ζ, we consider slowly varying function A and its derivatives as constants at the scale of the wave period, to obtain, instead of (1.32),

$$u_1 = \frac{1}{2k^2\left(V^2 - c_0^2\right)}\left(g\omega A + 2kc_0^2\frac{\partial A}{\partial X} + 2kV\frac{\partial A}{\partial T}\right)\left(k\zeta\cos k\zeta - \sin k\zeta\right). \quad (1.35)$$

Thus, to keep u_1 limited, the following equation must be satisfied:

$$\frac{\partial A}{\partial T} + \frac{c_0^2}{V}\frac{\partial A}{\partial X} + \frac{g\omega}{2kV}A = 0. \quad (1.36)$$

In this approximation $u_1 = 0$, and in the main order the wave amplitude slowly varies with time and coordinate according to (1.36). A general solution of (1.36) is

$$A = A_0 e^{-gT/2} F(X - c_g T), \quad (1.37)$$

and, therefore,

$$u = A_0 e^{-gT/2} F(X - c_g T)\cos\left(kx - \omega_0 t\right). \quad (1.38)$$

Here F is an arbitrary function depending on the boundary or initial condition. The latter expression is somewhat similar to (1.12) with the differences being that the wave frequency is arbitrary in the limits of $1 < \omega < \infty$, the carrier harmonic wave propagates with phase velocity $V = \omega/k$, and the wave amplitude propagates with group velocity, with a slow attenuation.

In general, even without damping ($g = 0$), the wave can be modulated in time and space, $A = A(X, T)$, due to the initial and/or boundary conditions. For a non-damped wave ($g = 0$) we have

$$\frac{\partial A}{\partial t} + c_g\frac{\partial A}{\partial x} = 0, \text{ so that } A = F(x - c_g t). \quad (1.39)$$

In higher approximations the form of such a wave train is not conserved.

1.2.2. *Nonlinear wave*

Now consider a modulated wave in the following nonlinear wave equation (the nonlinear Klein–Gordon equation):

$$\frac{\partial^2 u}{\partial t^2} - c_0^2 \frac{\partial^2 u}{\partial x^2} + \omega_0^2 u = -\mu q u^3. \tag{1.40}$$

We already know enough to pass over the formal perturbation scheme and look for the modulated solution in the form similar to (1.33), which will now be written in terms of complex amplitude but which still remains real (this is important for nonlinear equations written for a real function):

$$u = A(X,T)e^{i(kx-\omega t)} + A^*(X,T)e^{-i(kx-\omega t)} + \mu u_1(x,t)\dots, \tag{1.41}$$

where A^* is the complex conjugate of A. Substituting this into (1.40), at $\mu = 0$ we obtain the solution equivalent to (1.28),

$$u = Ae^{i(kx-\omega t)} + c.c.,\ k^2 = \omega_0^2 / \left(V^2 - c_0^2\right),$$

with the dispersion equation (1.29). Here and further, *c.c.* denotes a complex conjugated term.

For $\mu \neq 0$ we have

$$c_0^2 \frac{\partial^2 u_1}{\partial x^2} - \frac{\partial^2 u_1}{\partial t^2} - \omega_0^2 u_1 = \left(3qA^2A^* - 2ikc_0^2 \frac{\partial A}{\partial X} - 2i\omega \frac{\partial A}{\partial T}\right) e^{i(kx-\omega t)}$$
$$+ qA^3 e^{3i(kx-\omega t)} + c.c. \tag{1.42}$$

The "forcing" in the right-hand side of this equation contains both the first and the third harmonics. As above, we should be mainly concerned about the former, which acts in a resonance manner. Indeed, seeking the solution in the form

$$u_1 = ae^{i(kx-\omega t)} + be^{3i(kx-\omega t)} + c.c.,$$

we easily find that

$$a = \frac{S_a}{\omega^2 - c_0^2 k^2 - \omega_0^2}, \quad b = \frac{S_b}{9\omega^2 - 9c_0^2 k^2 - \omega_0^2}, \tag{1.43}$$

where S_a and S_b are, respectively, the coefficients at the first and the third harmonics in the right-hand side of (1.42). From the dispersion equation (1.29) we immediately see that the denominator in the first expression in (1.43) is zero, whereas the second term remains finite. To eliminate such a divergence, we have to let $S_a = 0$, or

$$\frac{\partial A}{\partial T} + c_g \frac{\partial A}{\partial X} = -\frac{3iq}{2\omega} |A|^2 A. \tag{1.44}$$

Here, $c_g = c_0^2 / V$ is again the linear group velocity. For a wave with a constant amplitude modulus, $A = A_0 e^{ik'X}$, where A_0 = const., from (1.44) we have

$$k' = \frac{3q}{c_g} |A_0|^2, \tag{1.45}$$

so that the full wave number, $k + \mu k'$, depends on the wave amplitude. This is a simple example of nonlinear "self-action": the wave parameter (in this case wavelength) depends on amplitude and, if the latter is modulated, the parameter can change in the course of propagation.

Now we briefly consider the next approximation in μ. Namely, in the resonance term in (1.42) we retain the second derivatives of the amplitude:

$$S_a = 3qA^2A^* - 2ikc_0^2 \frac{\partial A}{\partial X} - 2i\omega \frac{\partial A}{\partial T} + \frac{\partial^2 A}{\partial T^2} - c_0^2 \frac{\partial^2 A}{\partial X^2}.$$

To eliminate resonance, we must make the entire expression for S_a equal to zero; i.e.,

$$\frac{\partial A}{\partial T} + c_g \frac{\partial A}{\partial X} + i\beta \frac{\partial^2 A}{\partial X^2} + i\gamma |A|^2 A = 0, \tag{1.46}$$

where $\beta = (c_g^2 - c_0^2)/(2kc_0^2)$ and $\gamma = 3q/(2kc_0^2)$; in the small term with β we used the relation $\frac{\partial A}{\partial T} \approx -c_g \frac{\partial A}{\partial X}$ which is true up to the next order. Finally, using the variables T and $X' = X - c_g T$ instead of T and X in (1.46), we obtain

$$\frac{\partial A}{\partial T} + i\beta \frac{\partial^2 A}{\partial X'^2} + i\gamma |A|^2 A = 0. \tag{1.47}$$

Equation (1.47) is widely used in contemporary theory of nonlinear waves. It is called the *Nonlinear Schrödinger equation* (NSE or NLS). This equation has been thoroughly discussed in the literature, both mathematically and in applications to nonlinear optics, water waves, etc. (e.g., [2, 7–9]). The range of its solutions includes such important classes as modulation instability (a periodic wave close to a harmonic one can be unstable with respect to small modulation) and "envelope solitons", stationary wave packets propagating without deformation of its amplitude function. We shall return to some of these phenomena later in the book.

1.3. Concluding Remarks

The relatively simple examples discussed above demonstrate the specifics of the asymptotic perturbation approach. In the simple expansions with the fixed main term we often encounter the cases in which the supposedly small perturbation cumulatively (secularly) increases due to the resonant action of small terms in the equation. In the asymptotic schemes these perturbations are incorporated into the main term, thus allowing its parameters (such as amplitude) to slowly but possibly strongly vary in time and, for the waves, in space. In the next chapters we describe the asymptotic theory for the rather general classes of wave equations, and continue to consider various applications.

References

1. Andronov, A. A., Vitt, A. A., and Khaikin, S. E. (1966). *Theory of Oscillators*. Pergamon Press, Oxford.

2. Ablowitz, M. J. and Segur, H. (1981). *Solitons and the Inverse Scattering Transform.* SIAM, Philadelphia.
3. Bender, C. M. and Orszag, S. A. (1978). *Advanced Mathematical Methods for Scientists and Engineers: Asymptotic Methods and Perturbation Theory.* McGraw-Hill, New York.
4. Bogolyubov, N. N. and Mitropolsky, Y. A. (1961). *Asymptotic Methods in the Theory of Non-Linear Oscillations.* Gordon and Breach, New York.
5. Guckenheimer, J. and Holmes, P. (1983). *Nonlinear Oscillations, Dynamical Systems, and Bifurcations of Vector Fields.* Springer-Verlag, Berlin.
6. Korn, G. A. and Korn, T. (1968). *Mathematical Handbook.* McGraw-Hill, New York.
7. Ostrovsky, L. A. and Potapov, A. I. (1999). *Modulated Waves: Theory and Applications.* Johns Hopkins University Press, Baltimore and London.
8. Debnath, L. (2005). *Nonlinear Partial Differential Equations for Scientists and Engineers.* 2nd Ed. Birkhauser, Boston.
9. Malomed, B. (2005). Nonlinear Schrodinger equations, in *Encyclopedia of Nonlinear Science*, ed. A. Scott. Routledge, New York and London, pp. 639–643.
10. Van der Pol, B. (1927). On relaxation-oscillations, *The London, Edinburgh and Dublin Philosophical Magazine and Journal of Science*, v. 2, pp. 978–992.

Chapter 2

Perturbation Method for Quasi-Harmonic Waves

Make things as simple as possible, but not simpler.

Albert Einstein

Let us now apply the ideas outlined in the previous chapter to a rather general system of weakly nonlinear equations with dispersion which, in the absence of perturbations, possess a particular family of solutions in the form of plane harmonic waves. First we describe a general scheme and then give specific physical examples. Since the wave parameter variation can be one-, two-, or three-dimensional, we use slow time $T = \mu t$ and "slow" position vector $\mathbf{X} = \mu \mathbf{r}$, where, as above, $\mu \ll 1$ is a small parameter. Correspondingly, the wave parameters (such as amplitude and frequency) depend on slow variables, and the n-th derivatives of these parameters are of the order of μ^n.

2.1. General Scheme

Consider a set of N equations which at $\mu = 0$ are linear and have constant coefficients, so that nonlinear and other perturbing terms are small, of the order of μ or higher, whereas the coefficients of the system can be slowly varying:

$$A_{ik}(T,\mathbf{X})\frac{\partial u_k}{\partial t}+B_{ik}^{(s)}(T,\mathbf{X})\frac{\partial u_k}{\partial r_s}+C_{ik}(T,\mathbf{X})u_k=\mu f_i(u_l,T,\mathbf{X}),$$
$$i,k=1,...N,\ \ r_s=\{x,y,z\}, \tag{2.1}$$

or, in the vector-matrix form,

$$A(T,\mathbf{X})\frac{\partial \mathbf{u}}{\partial t}+\mathbf{B}(T,\mathbf{X})\frac{\partial \mathbf{u}}{\partial \mathbf{r}}+C(T,\mathbf{X})\mathbf{u}=\mu\mathbf{f}(\mathbf{u},T,\mathbf{X}). \tag{2.2}$$

Here $\mathbf{u}(\mathbf{r},t)=\{u_1,u_2,...u_N\}$ is a set (vector) of N unknown functions; $\mathbf{r}=\{x, y, z\}$ is the coordinate vector; A, $\mathbf{B}=\{B^{(x)},B^{(y)},B^{(z)}\}$, and C are square ($N \times N$) matrices of coefficients which may depend on slow variables T and $\mathbf{X}$. Note that in (2.2), $\partial u_i/\partial\mathbf{r}=\nabla\cdot u_i$ is the spatial gradient, so that $\mathbf{B}$ is the vector matrix having three components corresponding to the components of vector $\mathbf{u}$. The operator $\mathbf{f}$, which is in general nonlinear, can depend on $\mathbf{u}$, its derivatives, and, possibly, integrals.

As mentioned, we suppose that at $\mu = 0$ there exists a family of solutions in the form of a plane harmonic wave:

$$\mathbf{u}(\theta,\boldsymbol{a})=\boldsymbol{a}\exp(i\theta)+c.c.,\ \ \theta=\omega t-\mathbf{kr}. \tag{2.3}$$

Here, as above, *c.c.* denotes complex conjugation (we look for real solutions). Substitution of (2.3) into (2.2) at $\mu = 0$ generates a homogeneous, N-th order linear algebraic system for the components of the amplitude vector $\boldsymbol{a}$:

$$(A\omega-\mathbf{Bk}-iC)\boldsymbol{a}=0. \tag{2.4}$$

For a non-trivial solution of the latter to exist, the following condition (dispersion equation) should be met:

$$D(\omega,k)=Det\|A\omega-\mathbf{Bk}-iC\|=0. \tag{2.5}$$

At a given $\mathbf{k}$, this equation has, in general, N solutions for ω. For each solution, $\omega(\mathbf{k}_m)$ (a "normal mode" of a number $m=1, 2, \ldots, N$),

Eq. (2.4) defines the amplitude $\boldsymbol{a}_s$ as a right eigenvector of the matrix $||A\omega - \mathbf{Bk} - iC||$. In what follows we consider propagating waves, for which both ω and $\mathbf{k}$ are real, and omit the subscript m. Another important point: it is assumed that at $\mu = 0$, our system is stable, i.e., at real $\mathbf{k}$, the solution of (2.4) is such that Im$\omega \geq 0$ for all modes, and Imω = 0 for the mode considered.

Let us return to the perturbed system (2.2) with $\mu \neq 0$. We represent its solution as a series:

$$\mathbf{u} = \boldsymbol{a}(T, \mathbf{X})\exp(i\theta) + \mu\mathbf{u}^{(1)}(\theta, T, \mathbf{X}) + \mu^2\mathbf{u}^{(2)}(\theta, T, \mathbf{X}) + ... + c.c. \quad (2.6)$$

The wave frequency and wave number can now be slowly varying; namely,

$$\omega(T, \mathbf{X}) = \frac{\partial\theta}{\partial t}, \ \mathbf{k}(T, \mathbf{X}) = \nabla \cdot \theta = \frac{\partial\theta}{\partial \mathbf{r}}, \quad (2.7)$$

so that

$$\frac{\partial \mathbf{k}}{\partial t} + \frac{\partial\omega}{\partial \mathbf{r}} = 0 \,. \quad (2.8)$$

Any fixed branch of solutions of (2.4) defines a dependence

$$\omega = \omega(\mathbf{k}, T, \mathbf{X}), \quad (2.9)$$

i.e., the dispersion equation. Substituting this into (2.8), we have

$$\frac{\partial\omega}{\partial t} + \mathbf{c}_g \frac{\partial\omega}{\partial \mathbf{r}} = 0, \ \mathbf{c}_g = \frac{\partial\omega}{\partial \mathbf{k}}, \text{ so that } c_{gs} = \frac{\partial\omega}{\partial k_s} \ (s = x, y, z). \quad (2.10)$$

Here $\mathbf{c}_g$ is the group velocity vector.

After substitution of (2.6) into (2.2), in the main approximation we obtain Eqs (2.4) and (2.5). In the next order we have

$$A\frac{\partial \mathbf{u}_1}{\partial t} + \mathbf{B}\frac{\partial \mathbf{u}_1}{\partial \mathbf{r}} + C\mathbf{u}_1 = \mathbf{f} - \mathbf{H}^{(0)}, \ \mathbf{H}^{(0)} = \left(A\frac{\partial \boldsymbol{a}}{\partial T} + \mathbf{B}\frac{\partial \boldsymbol{a}}{\partial \mathbf{X}}\right)e^{i\theta}. \quad (2.11)$$

The operator $\mathbf{f}$ is in general periodic in θ and, hence, it can be represented by the Fourier series:

$$\mathbf{f} = \sum_n \mathbf{f}_n \exp(in\theta) + c.c., \quad n = 1,2,\dots . \tag{2.12}$$

Here $\mathbf{f}_n(T,\mathbf{X})$ are slowly varying coefficients. Substituting this into (2.11), we represent the perturbation $\mathbf{u}_1$ in the form of the Fourier series:

$$\mathbf{u}_1 = \sum_n \mathbf{u}_{1n} \exp(in\theta) + c.c., \tag{2.13}$$

where $\mathbf{u}_{1n}(T,\mathbf{X})$ are slowly varying Fourier coefficients. Subsequently the solution of (2.11) yields

$$\mathbf{u}_1 = \frac{\mathbf{f}_1 - \left(A\dfrac{\partial a}{\partial T} + \mathbf{B}\dfrac{\partial a}{\partial \mathbf{X}} \right)}{D(\omega,k)}, \quad \mathbf{u}_2 = \frac{\mathbf{f}_2}{D_2(\omega,k)} \dots . \tag{2.14}$$

Here $D(\omega,k)$ is given by (2.3) and

$$D_2(\omega,k) = Det\left\| 2A\omega - 2\mathbf{Bk} - iC \right\| \tag{2.15}$$

is the determinant obtained similarly to (2.3) when ω is replaced by 2ω (for higher harmonics it would be 3ω, 4ω, etc.).

According to (2.4), D = 0. First we suppose that D_2 and similar determinants for higher harmonics are non-zero for all T and $\mathbf{X}$; in the next section we shall consider the resonant interaction of wave harmonics for which the latter condition is violated.

As seen from (2.14), the first harmonic of perturbation diverges unless the condition

$$A\frac{\partial a}{\partial T} + \mathbf{B}\frac{\partial a}{\partial \mathbf{X}} = \mathbf{f}_1 \tag{2.16}$$

is met. This equation defines modulation of the complex amplitude of the wave in the leading-order approximation.

Differentiating (2.5) with respect to $\mathbf{k}$, we readily obtain

$$I\frac{\partial \omega}{\partial \mathbf{k}} = I\mathbf{c}_g = A^{-1}\mathbf{B},$$

where I is the unit matrix and A^{-1} is the matrix inverse to A. Hence, rewriting (2.16) in the form

$$\frac{\partial a}{\partial T} + A^{-1}\mathbf{B}\frac{\partial a}{\partial \mathbf{X}} \equiv \frac{\partial a}{\partial T} + I\mathbf{c}_\mathbf{g}\frac{\partial a}{\partial \mathbf{X}} = A^{-1}\mathbf{f}_1, \tag{2.17}$$

we conclude that at $\mathbf{f}_1 = 0$, the amplitude of each wave mode propagates with the corresponding group velocity.

These equations define slow variation of the wave amplitude. After that the non-resonant, higher-order perturbation remains finite and can be found subsequently; for example, $\mathbf{u}_2$ is defined by the second relation (2.14).

2.2. Resonant Interaction of Waves

In the expressions (2.14), only the resonance between the perturbation and a single wave mode with the phase θ was assumed, so that the other, non-resonant perturbation modes can be found by a regular, rather than asymptotic, procedure. There is, however, the possibility that more resonances do exist; for example, the denominator D_2 in (2.14) may turn to zero. In a general case, instead of (2.13), a multi-wave field should be sought:

$$\begin{aligned} &\mathbf{u} = \sum_{m,n} \boldsymbol{a}_m(T,\mathbf{X})\exp(i\theta_m) + \mu\mathbf{u}^{(1)}(\theta_1,\theta_2,...\theta_l) + ... + c.c.; \\ &m = 1,2,...l, \end{aligned} \tag{2.18}$$

with $\theta_m = \omega_m t - \mathbf{k}_m\mathbf{r}$. We assume that all l quasi-harmonic waves can propagate in an unperturbed linear system; i.e., each of them satisfies the condition (2.5). Their nonlinear interaction creates a new wave with the frequency ω_p and wave number $\mathbf{k}_\mathbf{p}$, so that

$$\sum_{m,n} n\omega_m = \omega_p, \ \sum_{m,n} n\mathbf{k}_m = \mathbf{k}_p, \ n = \pm 1, \pm 2, \ldots . \tag{2.19}$$

In the case when the new wave is also propagating, i.e. it also satisfies (2.5), it can resonantly interact with the others and should be included in the main approximation in (2.18).

A simple but important example is the three-wave interaction:

$$\omega_1 \pm \omega_2 = \omega_3, \quad \mathbf{k}_1 \pm \mathbf{k}_2 = \mathbf{k}_3, \tag{2.20}$$

or its particular case of frequency doubling or period doubling when $\omega_1 = \omega_2 = \omega$ and $\mathbf{k}_1 = \mathbf{k}_2 = \mathbf{k}$, so that $\omega_3 = 2\omega$.

Now, substituting (2.19) or (2.20) into (2.18), then into (2.2), and expanding $\mathbf{f}$ in the multiple Fourier series involving frequencies of all main harmonic components, we obtain the perturbation components similar to (2.14):

$$\mathbf{u}_m = \frac{\mathbf{f}_m - \left(A\dfrac{\partial \mathbf{a}_m}{\partial T} + \mathbf{B}\dfrac{\partial \mathbf{a}_m}{\partial \mathbf{X}} \right)}{D_m(\omega,\mathbf{k})}, \qquad D_m(\omega,\mathbf{k}) = Det\left\| A\omega_m - \mathbf{B}\mathbf{k}_m - C \right\|. \tag{2.21}$$

Here $m = 1,2,...l$ refer to resonant wave perturbations. To secure the finiteness of perturbations, a set of orthogonality conditions, which are the coupled equations of the type (2.16), should be imposed:

$$A\frac{\partial \mathbf{a}_m}{\partial T} + \mathbf{B}\frac{\partial \mathbf{a}_m}{\partial \mathbf{X}} = \mathbf{f}_m, \quad m = 1,2,...l. \tag{2.22}$$

A more detailed analysis of a similar scheme for a more complex equation set can be found in [8].

Below some physical examples illustrating the general schemes of the method are given.

2.3. Geometrical Acoustics

We begin from linear equations with slowly varying coefficients. The term "geometrical optics" or "ray optics" is widespread in literature. It means an approximate description of light or, more generally, electromagnetic waves, with a wavelength much smaller than the scale at which the wave parameters, such as amplitude and wavelength, significantly vary. Such waves can be commonly described in terms of "rays" along which the wave propagates. This representation becomes invalid in points where the rays intersect, such as foci and caustics, where the effect of diffraction must be taken into account. Naturally, the "geometrical" description can be used for waves of any physical origin. To be specific, here we consider acoustic waves in the framework of "geometrical acoustics".

Consider propagation of sound in a smoothly inhomogeneous medium. In the linear approximation, the corresponding equations have the form [2, 3, 9]:

$$\begin{aligned} &\frac{\partial \mathbf{v}}{\partial t}+\frac{\nabla p'}{\rho_0}-\frac{\nabla p_0}{\rho_0^2}\rho'=0, \\ &\frac{\partial \rho'}{\partial t}+\rho_0(\nabla\cdot\mathbf{v})+(\mathbf{v}\cdot\nabla)\rho_0=0, \\ &\frac{\partial p'}{\partial t}+\mathbf{v}\cdot\nabla p_0=c_0^2\left(\frac{\partial \rho'}{\partial t}+\mathbf{v}\cdot\nabla\rho_0\right). \end{aligned} \tag{2.23}$$

Here p_0, ρ_0, and $c_0=\sqrt{dp_0/d\rho_0}$ are, respectively, slowly varying background pressure, density, and sound speed; $\mathbf{v}$, p', and ρ' are particle velocity, pressure, and density perturbations in the acoustic wave.

To make the equations less cumbersome, consider two-dimensional propagation of sound in the (x, y) plane, assuming that p_0 and ρ_0 are functions of x and y, and the vector $\mathbf{v}$ has two components, $v_x = u$ and $v_y = w$. The three-dimensional generalization is straightforward. In the two-dimensional case, Eqs (2.23) are reduced to

$$\begin{aligned} &\rho_0\frac{\partial u}{\partial t}+\frac{\partial p'}{\partial x}=\mu\frac{\partial p_0}{\partial X}\frac{\rho'}{\rho_0}, \\ &\rho_0\frac{\partial w}{\partial t}+\frac{\partial p'}{\partial y}=\mu\frac{\partial p_0}{\partial Y}\frac{\rho'}{\rho_0}, \\ &\frac{\partial \rho'}{\partial t}+\rho_0\left(\frac{\partial u}{\partial x}+\frac{\partial w}{\partial y}\right)=-\mu\left(u\frac{\partial \rho_0}{\partial X}+w\frac{\partial \rho_0}{\partial Y}\right), \\ &\frac{\partial p'}{\partial t}-c_0^2\frac{\partial \rho'}{\partial t}=\mu u\left(-\frac{\partial p_0}{\partial X}+c_0^2\frac{\partial \rho_0}{\partial X}\right)+\mu w\left(-\frac{\partial p_0}{\partial Y}+c_0^2\frac{\partial \rho_0}{\partial Y}\right). \end{aligned} \tag{2.24}$$

Here, again μ is a small parameter, and $X=\mu x$ and $Y=\mu y$ are slow variables. Representing these equations in the vector-matrix form corresponding to (2.2), we denote

$$\mathbf{u}=\begin{Bmatrix} u_x \\ u_y \\ p' \\ \rho' \end{Bmatrix},\ A=\begin{Bmatrix} \rho_0 & 0 & 0 & 0 \\ 0 & \rho_0 & 0 & 0 \\ 0 & 0 & 1 & 0 \\ 0 & 0 & -c_0^2 & 1 \end{Bmatrix},\ B^{(x)}=\begin{Bmatrix} 0 & 0 & 0 & 1 \\ 0 & 0 & 0 & 0 \\ \rho_0 & 0 & 0 & 0 \\ 0 & 0 & 0 & 0 \end{Bmatrix},$$

$$B^{(y)}=\begin{Bmatrix} 0 & 0 & 0 & 0 \\ 0 & 0 & 0 & 1 \\ 0 & \rho_0 & 0 & 0 \\ 0 & 0 & 0 & 0 \end{Bmatrix},\quad \mathbf{f}=\begin{Bmatrix} \dfrac{\partial\rho_0}{\partial X}\dfrac{\rho'}{\rho_0} \\ \dfrac{\partial\rho_0}{\partial Y}\dfrac{\rho'}{\rho_0} \\ v_x\dfrac{\partial\rho_0}{\partial X}+v_y\dfrac{\partial\rho_0}{\partial Y} \\ v_x\left(-\dfrac{\partial p_0}{\partial X}+c_0^2\dfrac{\partial\rho_0}{\partial Y}\right) \\ +v_y\left(-\dfrac{\partial p_0}{\partial X}+c_0^2\dfrac{\partial\rho_0}{\partial Y}\right) \end{Bmatrix}, \tag{2.25}$$

$$C=0.$$

According to (2.3), consider the quasi-harmonic wave whose amplitude is slowly varying in space and time, and represent the components of the vector **u** in the form:

$$\begin{aligned} u&=\left[U(X,Y,T)+\mu u_1\right]\exp(i\theta),\ w=\left[W(X,Y,T)+\mu w_1\right]\exp(i\theta), \\ \rho'&=\left[R(X,Y,T)+\mu\rho_1\right]\exp(i\theta),\ p'=\left[P(X,Y,T)+\mu p_1\right]\exp(i\theta), \end{aligned} \tag{2.26}$$

where $X=\mu x$ and $Y=\mu y$ are slow coordinates, and $T=\mu t$ is slow time. The parameters ρ_0 and c_0 are, in general, functions of X and Y but not of T, i.e., the medium is supposed stationary.

In the main order, the dispersion equation (2.4) now reads

$$\omega^2 = c_0^2\left(k_x^2 + k_y^2\right) = c_0^2 k^2, \tag{2.27}$$

and from the algebraic equations (2.4) it follows that

$$\begin{aligned} &P = c_0^2 R = \rho c_0 a;\;\; a = \sqrt{U^2 + W^2}, \\ &\frac{U}{W} = \frac{k_x}{k_y},\;\; \{U, W\} = \{k_x, k_y\}\frac{a}{k},\;\; k = \sqrt{k_x^2 + k_y^2}. \end{aligned} \tag{2.28}$$

Here a is the modulus of the velocity vector $\mathbf{v}$ which is parallel to the wave vector $\mathbf{k}$. Locally, all this corresponds to the features of a plane acoustic wave in a homogeneous medium.

In the next-order approximation, we obtain equations of the (2.11) type; for quasi-harmonic perturbations it yields

$$\begin{aligned} &i\omega\rho_0 u_1 - ik_x p_1 = \frac{R}{\rho_0}\frac{\partial p_0}{\partial X} - \frac{\partial P}{\partial X} - \rho_0\frac{\partial U}{\partial T} = Q_{1x}, \\ &i\omega\rho_0 w_1 - ik_y p_1 = \frac{R}{\rho_0}\frac{\partial p_0}{\partial Y} - \frac{\partial P}{\partial Y} - \rho_0\frac{\partial W}{\partial T} = Q_{1y}, \\ &i\omega\rho_1 - i\rho_0\left(k_x u_1 + k_y w_1\right) = -U\frac{\partial \rho_0}{\partial X} - W\frac{\partial \rho_0}{\partial Y} - \rho_0\left(\frac{\partial U}{\partial X} + \frac{\partial W}{\partial Y}\right) - \frac{\partial R}{\partial T} = Q_2, \\ &i\omega(p_1 - c_0^2\rho_1) = -U\left(\frac{\partial p_0}{\partial X} - c_0^2\frac{\partial \rho_0}{\partial X}\right) - W\left(\frac{\partial p_0}{\partial Y} - c_0^2\frac{\partial \rho_0}{\partial Y}\right) - \left(\frac{\partial P}{\partial T} - c_0^2\frac{\partial R}{\partial T}\right) = Q_3. \end{aligned} \tag{2.29}$$

In these equations the relations (2.27) and (2.28) can be used in the same approximation. As a result we obtain

$$\omega Q_3 + c_0^2\left(k_x Q_{1x} + k_y Q_{1y} + \omega Q_2\right) = 0. \tag{2.30}$$

After substitution of the expressions for $Q_{1,2,3}$ from (2.29) and a few elementary transformations using (2.28) we obtain

$$\omega\left(\frac{\partial P}{\partial T}\right)+c_0^2\rho_0\left(k_x\frac{\partial U}{\partial T}+k_y\frac{\partial W}{\partial T}\right)+c_0^2\left(k_x\frac{\partial P}{\partial X}+k_y\frac{\partial P}{\partial Y}\right)$$
$$+c_0^2\rho_0\omega\left(\frac{\partial U}{\partial X}+\frac{\partial W}{\partial Y}\right)=0. \tag{2.31}$$

Substituting the relations (2.28), we arrive, in the same approximation, to one equation for, e.g., the velocity modulus, *a*:

$$2\omega\rho_0 c_0\frac{\partial a}{\partial T}+2c_0^3\rho_0(\mathbf{k}\nabla)a+c_0^2 a\left[(\mathbf{k}\nabla)(c_0\rho_0)+\rho_0\omega\nabla(\mathbf{k}/k)\right]=0. \tag{2.32}$$

Here ∇ is the gradient operator. The variables *U*, *W*, and *P* are related to *a* via (2.28).

Equations (2.27) and (2.32) allow a simple interpretation as those of geometrical acoustics [3]. Indeed, (2.27) is equivalent to the equation for the phase θ (which is called *eikonal*, from the Greek "image"):

$$\left(\frac{\partial\theta}{\partial x}\right)^2+\left(\frac{\partial\theta}{\partial y}\right)^2=(\nabla\theta)^2=\frac{1}{c_0^2}\left(\frac{\partial\theta}{\partial t}\right)^2=\frac{\omega^2}{c_0^2}. \tag{2.33}$$

When the medium is stationary, i.e., c_0 and ρ_0 do not depend on *T*, the lines $\theta(x, y)$ = const. define the wave fronts, and normal to them are *rays*, which are the lines collinear to the wave vector **k** in each point. In a non-stationary medium, Eq. (2.33) defines the space-time rays in the (x, y, t) space.

Equation (2.32) expresses energy conservation. After multiplying by *a* it can be reduced to the form

$$\frac{\partial E}{\partial T}+(\nabla\cdot\mathbf{S})=0. \tag{2.34}$$

Here $E=\rho a^2$ and $\mathbf{S}=(\mathbf{k}/k)c_0 E$ are, respectively, the densities of acoustic energy and acoustic energy flux; the latter is directed along the rays. It is easy to verify that (2.34) is equivalent to (2.32). Similar results can be obtained in the three-dimensional case; actually, Eqs (2.32), (2.34), and, with adding $(\partial\theta/\partial z)^2$, (2.33) have exactly the same form.

Equations of geometrical acoustics and especially optics are thoroughly discussed in the literature. In particular, the *eikonal* equation (2.33) has the form of the Hamilton–Jacobi equation, which implies an analogy with mechanics of material particles [4].

For practical purposes it is convenient to represent gradients in Eq. (2.32) as derivatives along the direction $\boldsymbol{l}$ parallel to the vectors **k** and **v.** Note that the second term in square parentheses contains the derivative of a unit vector in the direction of **k**. It is easy to see that the rotation of this vector is equivalent to change of the width Q of the "ray tube" upon moving along l (Fig. 2.1).

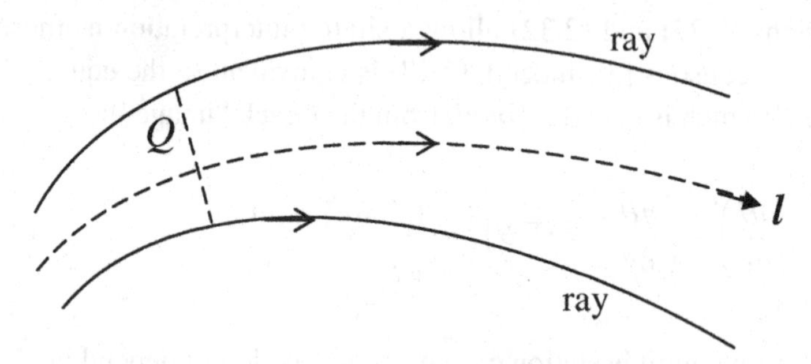

Figure 2.1. Variation of the width Q of the ray tube along the propagation direction $\boldsymbol{l}$.

As a result, (2.32) can be written in the scalar form:

$$\frac{\partial a}{\partial T}+c_0\frac{\partial a}{\partial l}+\frac{a}{2\rho_0 Q}\frac{\partial(c_0\rho_0 Q)}{\partial l}=0. \tag{2.35}$$

The solution of this equation has the form

$$a=(c_0\rho_0 Q)^{-1/2}F\left(T-\int_{l_0}^{l}dl'/c_0(l')\right), \tag{2.36}$$

where F is an arbitrary function to be found from a given boundary condition $a(l_0) = a_0(T)$, and l_0 is an initial point on the ray. If a is independent of T, i.e., the wave amplitude is not modulated in time, the solution is

$$a(l) = a_0\sqrt{\frac{f(l_0)}{f(l)}},\ f = c_0\rho_0 Q. \tag{2.37}$$

The shape of rays and, as a result, slow variation of the parameters c_0, ρ_0, and Q along the rays, can be found from the eikonal equation (2.33). If, for example, the medium parameters are constant and the wave fronts are concentric so that the rays are the radii which diverge from or converge to a central point (a circular wave), then $l = r$ and the solution (2.37) is $a = a_0\sqrt{r_0/r}$. This solution is valid far from the center, where $kr >> 1$; thus, in the converging wave ($r < r_0$) the amplitude cannot increase unlimitedly: near the center, nonlinear effects such as the shock wave formation become significant.

In the three-dimensional configuration, the ray tube cross section Q is two-dimensional, and in a symmetric spherical wave, $a = a_0(r_0/r)$, where r is the distance from the center and r is much larger than the wavelength.

Another example refers to a harmonic plane wave propagating along the x direction in a medium the parameters of which vary only along x. If at the initial point $x = 0$, harmonic oscillations are given, we have in this case $\partial/\partial T = 0$ and $l = X$, and from (2.32) it follows that

$$a(x) = a(0)\sqrt{\frac{c_0(0)\rho_0(0)}{c_0(X)\rho_0(X)}},\ R(x) = R(0)\sqrt{\frac{\rho_0(X)}{\rho_0(0)}}\left(\frac{c_0(0)}{c_0(X)}\right)^{3/2},$$
$$P = P(0)\sqrt{\frac{c_0(X)\rho_0(X)}{c_0(0)\rho_0(0)}}. \tag{2.38}$$

One of the applications of these formulae refers to upward propagation of infrasound generated in the atmosphere, for example, by earthquakes. Such a wave propagates from the ground upward in the atmosphere and can reach even the ionosphere, i.e., altitudes of 100 km and more. For the idealized model of the isothermal atmosphere, in which $c_0(X)$ = const. and $\rho_0(X)$ is proportional to $\exp(-X/H)$, where X is directed upward and H is the effective height (about 8.5 km), (2.38) yields [6]

$$\frac{a(X)}{a(0)} = \exp(X/2H), \quad \frac{R(X)}{R(0)} = \frac{P(X)}{P(0)} = \exp(-X/2H). \qquad (2.39)$$

Here the particle velocity (and hence, the displacement) increases upward exponentially, until the nonlinearity becomes important and the shock waves form, which results in the wave attenuation [6]. In reality the amplitudes of the velocity can reach dozens of m/s; such motions in the ionosphere were observed using radars (e.g., [5]).

Figure 2.2 illustrates the exponential increase of particle velocity amplitude (and, correspondingly, the vertical displacement amplitude) in a wave propagating upward.

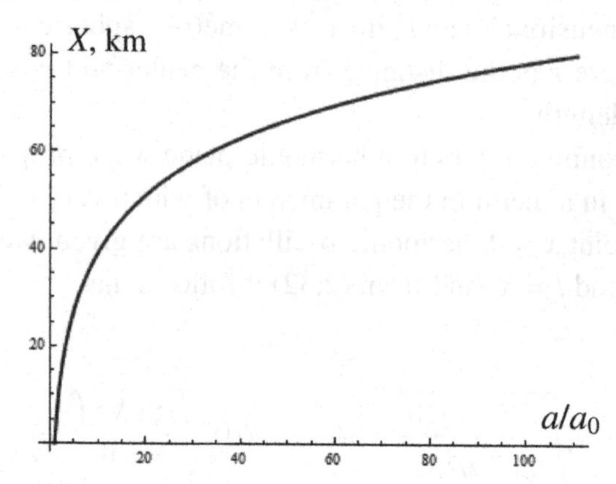

Figure 2.2. Dependence of normalized amplitude of the particle velocity oscillations (a/a_0) in the wave on vertical coordinate X at vertical propagation in isothermal atmosphere with H = 8.5 km.

Let us also mention propagation of an acoustic impulse which is short compared with the characteristic scale of spatial variation of medium parameters. Since the wave frequency and the pulse duration in a stationary non-dispersive medium remain constant, the wave number k is proportional to c_0 along the propagation direction l and, since the entire pulse is short, it is possible to neglect variations of c_0 and ρ at its scale. Then, integrating (2.34) along the propagation direction (ray) l, we obtain

$$\frac{d}{dT}\int_{-\infty}^{\infty} E dl = 0, \text{ or } W = \int_{-\infty}^{\infty} \rho U^2 dl = \text{const.}, \tag{2.40}$$

where W is the total wave energy. Integration is made across the entire pulse, formally at the whole l-axis.

2.4. Nonlinear Electromagnetic Waves in a Dispersive Dielectric

As an example of a nonlinear wave, we consider propagation of a plane, transversely polarized electromagnetic wave in a model of a dielectric consisting of dipoles with bounded electrons:

$$\begin{gathered}\frac{\partial E}{\partial x} + \mu_0 \frac{\partial H}{\partial t} = 0, \quad \frac{\partial H}{\partial x} + \varepsilon_0 \frac{\partial E}{\partial t} + \frac{\partial P}{\partial t} = 0, \\ \frac{\partial^2 P}{\partial t^2} + \omega_0^2 P - \omega_p^2 \varepsilon_0 E = \mu b P^3.\end{gathered} \tag{2.41}$$

Here E, H, and P are, respectively, transverse components of the electric and magnetic fields, and the medium polarization. The parameters ε_0 and μ_0 are, respectively, the electric permittivity and magnetic permeability of the vacuum (not to be confused with the small parameter μ). The parameters ω_0 and ω_p are, respectively, the resonance frequency of electron osscilations in an atom and the plasma frequency (the latter is proportional to the concentration of electrons); b is a phenomenological nonlinearity parameter (here nonlinearity is due to the non-parabolic character of the electron potential energy).

In the linear case, the last equation (2.41) is referred to as the Lorentz model. Nonlinear equations of this kind are widely used in electrodynamics and optics as simple models of nonlinear dispersion [7].

Strictly speaking, these equations are good for a rarefied dielectric (gas, plasma); otherwise, the electric field E in the last equation should be changed to some average field E_{eff} acting on a given electron from all other particles in the medium. The latter can be treated in a similar fashion.

After the simple substitution $\partial P/\partial t = S$, the system (2.41) reduces to the form (2.2). Again, we seek a quasi-harmonic solution:

$$H = [H_0(T,X) + \mu H_1]\exp(i\theta) + c.c., E = [E_0(T,X) + \mu E_1]\exp(i\theta) + c.c,$$
$$P = [P_0(T,X) + \mu P_1]\exp(i\theta) + c.c., \quad S(T,X) = [S_0 + \mu S_1]\exp(i\theta) + c.c., \tag{2.42}$$

and write the corresponding vectors and matrices of coefficients:

$$\mathbf{u} = \begin{Bmatrix} H \\ E \\ P \\ S \end{Bmatrix}, \quad A = \begin{vmatrix} \mu_0 & 0 & 0 & 0 \\ 0 & \varepsilon_0 & 1 & 0 \\ 0 & 0 & 0 & 1 \\ 0 & 0 & 1 & 0 \end{vmatrix}, \quad \mathrm{B} = \begin{vmatrix} 0 & 1 & 0 & 0 \\ 1 & 0 & 0 & 0 \\ 0 & 0 & 0 & 0 \\ 0 & 0 & 0 & 0 \end{vmatrix},$$

$$\mathrm{C} = \begin{vmatrix} 0 & 0 & 0 & 0 \\ 0 & 0 & 0 & 0 \\ 0 & -\omega_p^2\varepsilon_0 & \omega_0^2, & 0 \\ 0 & 0 & 0 & -1 \end{vmatrix}, \quad \mathbf{f} = \begin{Bmatrix} \mu_0 \dfrac{\partial H_0}{\partial T} + \dfrac{\partial E_0}{\partial X} \\ \varepsilon_0 \dfrac{\partial E_0}{\partial X} + \dfrac{\partial H_0}{\partial T} \\ \dfrac{\partial S_0}{\partial T} - bP_0^3 \\ \dfrac{\partial P_0}{\partial T} \end{Bmatrix}. \tag{2.43}$$

In particular, for the non-modulated harmonic wave at $\mathbf{f} = 0$, we obtain the dispersion equation (2.5) in the form:

$$c_0^2 k^2 = \omega^2 \left(\frac{\omega_0^2 + \omega_p^2 - \omega^2}{\omega_0^2 - \omega^2} \right), \tag{2.44}$$

where $c = \sqrt{\varepsilon_0 \mu_0}$ is the light velocity in vacuum. In this approximation all amplitudes can be expressed in terms of, e.g., E_0:

$$H_0 = \frac{k}{\omega \mu_0} E_0, \; P_0 = \frac{S_0}{i\omega} = \frac{\varepsilon_0 \omega_p^2}{\omega_0^2 - \omega^2} E_0. \quad (2.45)$$

In the next order we have

$$\begin{aligned}
& i\omega\mu_0 H_1 - ikE_1 = -\mu_0 \frac{\partial H_0}{\partial T} - \frac{\partial E_0}{\partial X} = q_1, \\
& i\varepsilon_0 \omega E_1 + i\omega P_1 - ikH_1 = -\varepsilon_0 \frac{\partial E_0}{\partial T} - \frac{\partial P_0}{\partial T} - \frac{\partial H_0}{\partial X} = q_2, \\
& i\omega S_1 + \omega_0^2 P_1 - \omega_p^2 \varepsilon_0 E_1 = 3bP_0^2 P_0^* - \frac{\partial S_0}{\partial T} = q_3, \\
& i\omega P_1 - S_1 = -\frac{\partial P_0}{\partial T} = q_4.
\end{aligned} \quad (2.46)$$

The linear algebraic system (2.46) has a non-trivial solution if

$$(k^2 - \varepsilon_0 \mu_0 \omega^2)(q_3 \omega + i q_4 \omega_0^2) + i\varepsilon_0 \omega \omega_p^2 \left[kq_1 + \mu_0 \omega (q_2 - q_4) \right] = 0. \quad (2.47)$$

Substituting here the expressions for q_i from (2.46) and the relations (2.45), we obtain

$$\frac{\varepsilon_0 \mu_0 \omega^4 - k^2 \omega_0^2}{\omega^2 - \omega_0^2} \frac{\partial E_0}{\partial T} + k\omega \frac{\partial E_0}{\partial X} = \frac{3ib\varepsilon_0^2 \omega \omega_p^4}{2} \frac{\left(k^2 - \varepsilon_0 \mu_0 \omega^2\right)}{\left(\omega^2 - \omega_0^2\right)^3} |E_0|^2 E_0, \quad (2.48)$$

or, after substituting the dispersion equation (2.44),

$$\frac{\partial E_0}{\partial T} + c_g \frac{\partial E_0}{\partial X} = \frac{3ib\varepsilon_0^2 \omega \omega_p^6 |E_0|^2 E_0}{2\left(\omega^2 - \omega_0^2\right)^2 \left[\left(\omega^2 - \omega_0^2\right)^2 - \omega_0^2 \omega_p^2 \right]} = ig|E_0|^2 E_0. \quad (2.49)$$

Here,

$$c_g = c_0 \frac{\left(\omega^2-\omega_0^2\right)^{3/2}\sqrt{\left(\omega^2-\omega_0^2-\omega_p^2\right)}}{\left[\left(\omega^2-\omega_0^2\right)^2+\omega_0^2\omega_p^2\right]} \tag{2.50}$$

is the group velocity of a wave packet, and

$$g = \frac{3b\varepsilon_0^2\omega\omega_p^4\left(k^2-\varepsilon_0\mu_0\omega^2\right)}{\left(\omega^2-\omega_0^2\right)^2\left[\left(\left(\omega^2-\omega_0^2\right)^2-\omega_0^2\omega_p^2\right)\right]} \tag{2.51}$$

can be considered as the nonlinearity parameter.

Dependencies of the phase and group velocities on the wave frequency are shown in Fig. 2.3.

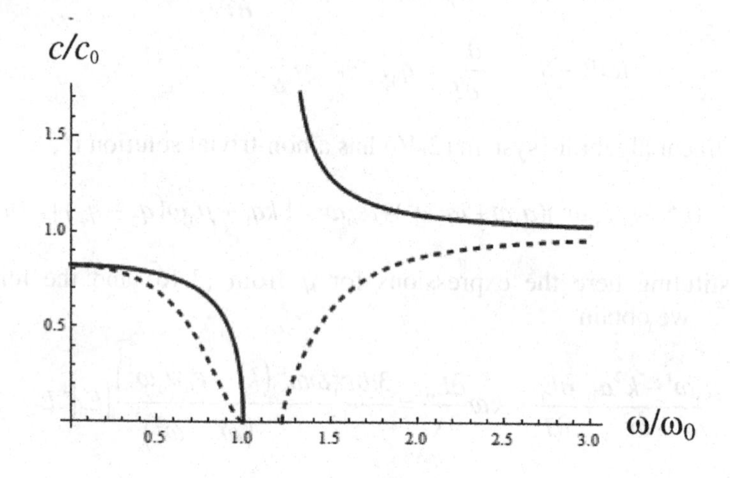

Figure 2.3. Phase velocity $c_p = \omega/k$ and group velocity $c_g = d\omega/dk$ normalized to c_0 as the functions of frequency according to (2.44) and (2.50), plotted for $\omega_p/\omega_0 = 0.7$. Solid line: c_p. Dashed line: c_g.

In the interval $\omega_0 < \omega < \sqrt{\omega_0^2+\omega_p^2}$, there is no real k in the dispersion equation (2.44) (non-transparency region). At large ω, both phase and group velocities tend to the light velocity in vacuum, c_0. The latter is understandable: the extremely short waves do not perturb the atomic structure of a material. However, this is a very high limit: even such short-wave radiation as the gamma-rays having frequencies of the order of 10^{19} Hz and wavelengths of the order of 10^{-11} m are absorbed by the condensed matter at distances of the order of 1 cm (which, however, amounts to 10^9 wavelengths) due to various mechanisms of interaction with the atomic structure.

Note also that at $\omega_0 = 0$ (free electrons) we have a case of collisionless plasma, with the following dispersion equation and the relations between amplitudes:

$$\omega^2 = c_0^2 k^2 + \omega_p^2, \ H_0 = \frac{k}{\omega \mu_0} E_0, \ P_0 = 0. \tag{2.52}$$

Equation (2.49) is similar to (1.44) of the first chapter; thus, it can be deduced that in the next approximation, the nonlinear Schrödinger equation (1.47) would follow.

2.4.1. *Resonance triplet*

Finally, we briefly consider the case when three waves interact in resonance when the solution is represented by (2.18) with $l = 3$ and the frequencies and wave vectors of these three waves are related by (2.20). This is an important problem in nonlinear optics [1] and numerous other areas of physics. In the one-dimensional case considered here, the wave vectors are all directed along x and represented by scalars.

Note that the relations (2.20) have a simple geometrical interpretation: the vectors with the components (k_m, ω_m) on the (k, ω) plane must sum up as shown in Fig. 2.4. For the amplitudes of these three waves, Eqs (2.22) have the form

$$\begin{aligned}
&\frac{\partial a_1}{\partial T} + c_{g1} \frac{\partial a_1}{\partial X} = i w_1 a_3 a_2^*, \\
&\frac{\partial a_2}{\partial T} + c_{g2} \frac{\partial a_2}{\partial X} = i w_2 a_3 a_1^*, \\
&\frac{\partial a_3}{\partial T} + c_{g3} \frac{\partial a_3}{\partial X} = i w_3 a_1 a_2.
\end{aligned} \tag{2.53}$$

Here, $a_{1,2,3}$ are amplitudes of the three interacting waves (e.g., of the electric field strengths), $w_{1,2,3}$ are interaction coefficients proportional to g in (2.49), and c_{gs} are the group velocities (2.50) in which the values of $\omega_s = \omega_{1,2,3}$ are, respectively, substituted. The interaction coefficients w are defined from equations similar to (2.46) written for each wave, with the only difference being in q_3 where instead of $|P_0|^2 P_0^*$ there will be $P_{01}P_{02}$ in the equation for the third wave, and $P_{03}P_{1,2}^*$ for the other two waves. Accordingly, from (2.47) we obtain

$$w_3 = (k_3^2 - \omega_3^2 / c_0^2)\omega_3; \; w_{1,2} = (k_{1,2}^2 - \omega_{1,2}^2 / c_0^2)\omega_{1,2}. \tag{2.54}$$

Here k_s can be expressed via ω from the dispersion equation (2.44).

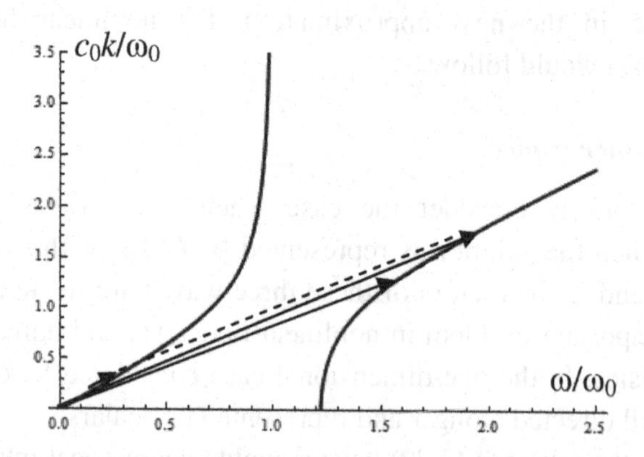

Figure 2.4. Dispersion curves $k(\omega)$ corresponding to (2.44) and three resonantly related waves with frequencies $\omega_1 = 0.44\omega_0$, $\omega_2 = 1.56\omega_0$, and $\omega_3 = 2\omega_0$. Solid straight lines show the vectors with components (ω, k); dashed lines illustrate the vector sum.

Equations of type (2.53) are widely known; their solution can be expressed in elliptic functions. Here we only mention that in this case all three waves cannot increase: some of them consume energy from the others. Indeed, if we multiply Eqs (2.53) by, respectively, a_1^*, a_2^*, and a_3^*, add them together with their complex conjugated, and use the resonance relations (2.20), it can be shown that

$$\frac{\partial}{\partial T}(W_1+W_2+W_3)+\frac{\partial}{\partial X}(S_1+S_2+S_3)=0,$$
$$W_m=\mu H_{0m}^2,\ S_m=c_{gm}W_m,\ m=1,2,3. \tag{2.55}$$

Here W_m is energy density in the m-th wave and S_m is its energy flux (note that in the linear approximation, average magnetic and electric energies in a plane traveling wave are equal).

It is seen from (2.55) that in the stationary case ($\partial/\partial T=0$) the total energy flux is conserved. Figure 2.5 shows how the energy periodically transforms between the third wave (the "pump wave") and two others.

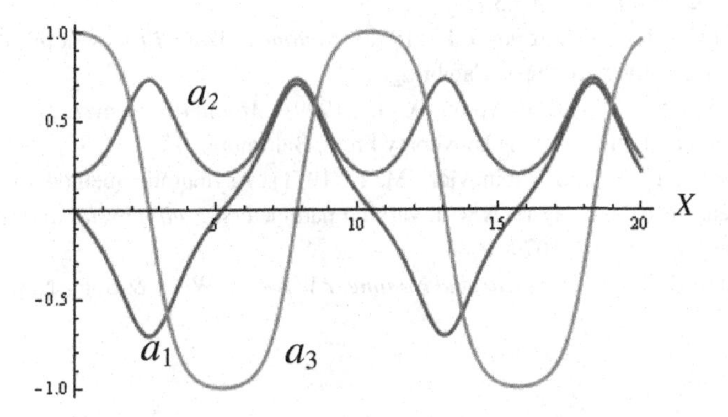

Figure 2.5. Variation of the waves' amplitudes along x with the third wave having the maximal frequency (arbitrary units).

2.5. Concluding Remarks

The asymptotic scheme for quasi-harmonic waves developed here for a rather general system (2.2) is relevant to a broad area of applications, including linear and nonlinear geometrical optics, geometrical acoustics, and others. In particular cases, various simplified approaches can be used, such as the ray theory, the energy balance method, etc. Nevertheless, the idea of the method is essentially the same. Many examples can be found elsewhere, so that we limit ourselves to the few examples discussed above, and move to the approximate description of the yet more general case of non-harmonic waves.

References

1. Bloembergen, N. (1996). *Nonlinear Optics.* 4th Ed. World Scientific, Singapore.
2. Brekhovskikh, L. M. (1980). *Waves in Layered Media.* Academic Press, New York and London.
3. Brekhovskikh, L. M. and Godin O. A. (1998). *Acoustics of Layered Media, I.* Springer, Berlin.
4. Landau, L. D. and Lifshits, E. M. (1987). *Fluid Mechanics.* Pergamon Press, Oxford and New York.
5. Liu, J. Y., Tsai, Y. B., Chen, S. W., Lee, C. P., Chen, Y. C., Yen, H. Y., Chang, W. Y., and Liu, C. (2006). Giant ionospheric disturbances excited by the M9.3 Sumatra earthquake of 26 December 2004, *Geophysiscal Research Letters* v. 33, L02103, doi:10.1029/ 2005GL023963.
6. Naugolnykh, K. and Ostrovsky, L. (1998). *Nonlinear Wave Processes in Acoustics.* Cambridge University Press, Cambridge.
7. Ostrovsky, L. A. and Potapov, A. I. (1999). *Modulated waves: Theory and Applications*. Johns Hopkins University Press, Baltimore.
8. Pelinovsky, E. N. and Rabinovich, M. I. (1971). Asymptotic method for weakly nonlinear, distributed systems with varying parameters, *Radiophysics and Quantum Electronics*, v. 14, pp. 1079–1086.
9. Whitham, G. B (1974). *Linear and Nonlinear Waves.* J. Wiley & Sons, New York.

Chapter 3

Perturbation Method for Non-Sinusoidal Waves

Absence of proof is not proof of absence.

Michael Crichton

3.1. General Scheme

The next generalization of the asymptotic perturbation method will involve periodic but not necessarily harmonic basic solutions. Now the unperturbed equations are already nonlinear, possibly strongly nonlinear. In general, the stationary wave profile is formed as a balance between the effects of nonlinearity and dispersion, and it can vary strongly depending on the wave amplitude. One example is shown in Fig. 3.1.

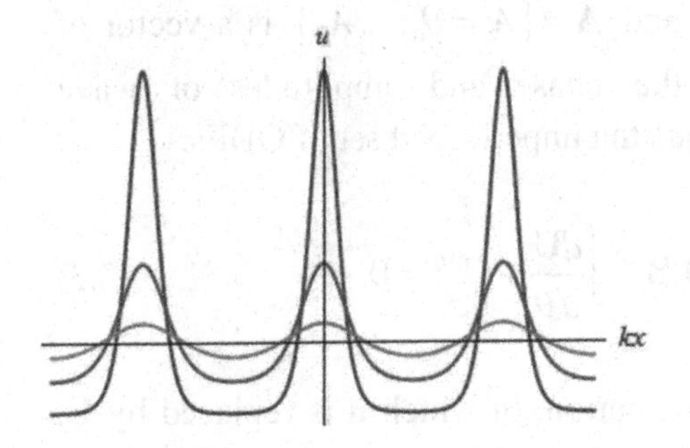

Figure 3.1. Profiles of steady progressive waves of different amplitudes with a given spatial period in the Korteweg–de Vries equation considered later in the book.

The idea of the asymptotic scheme remains the same as in Chapter 2, but its realization can be more cumbersome. Here we consider a set of N equations which differ from (2.2) in that at $\mu = 0$ they are nonlinear. For simplicity, these equations are supposed to be quasi-linear; i.e., their coefficients can depend on the unknown variable $\mathbf{u}$ but the unperturbed equations remain linear with respect to the derivatives of $\mathbf{u}$:

$$\begin{aligned} \mathbf{M}(\mathbf{u},T,\mathbf{X}) &= A(T,\mathbf{X},\mathbf{u})\frac{\partial \mathbf{u}}{\partial t} + \mathbf{B}(T,\mathbf{X},\mathbf{u})\frac{\partial \mathbf{u}}{\partial \mathbf{r}} + C(T,\mathbf{X},\mathbf{u}) \\ &= \mu \mathbf{f}(\mathbf{u},T,\mathbf{X}). \end{aligned} \tag{3.1}$$

Here, the notations are the same as in (2.2). The matrix coefficients A, $\mathbf{B} = \{B^{(x)}, B^{(y)}, B^{(z)}\}$, and C are square ($N \times N$) matrices which now depend on $\mathbf{u}$, and $\mathbf{f}$ is, in general, an operator which can depend on $\mathbf{u}$ and its derivatives. Note that in the original works [3, 7] the operator $\mathbf{M}$ is not necessarily quasi-linear, but here a somewhat more visual form is chosen.

We suppose that at $\mu = 0$ this system has a solution in the form of a periodic, and in general non-harmonic, plane wave:

$$\mathbf{u}(\mathbf{r},t) = \mathbf{U}(\theta,\mathbf{A}), \quad \theta = \omega t - (\mathbf{k}\cdot\mathbf{r}) + \theta_0, \tag{3.2}$$

where ω and $\mathbf{k}$ are the main frequency and wave vector determining the wave periods in time and space, and $\mathbf{A} = \{A_1 = \theta_0, \ldots, A_m\}$ is a vector of constant parameters (such as the phase and amplitudes of wave components). This solution satisfies the unperturbed set of ODEs,

$$\left[\omega A^{(0)} - \mathbf{k}\mathbf{B}^{(0)}\right]\frac{d\mathbf{U}}{d\theta} + C^{(0)} = 0. \tag{3.3}$$

Superscript 0 denotes zero approximation, in which $\mathbf{u}$ is replaced by $\mathbf{U}$. In periodic solutions, ω and $\mathbf{k}$ are related by a nonlinear dispersion equation: $\omega = \omega(\mathbf{k},\mathbf{A})$, where the dependence on $\mathbf{A}$ is due to the nonlinearity of the system (3.1).

For a perturbed system, at $\mu \neq 0$, the solution is represented by the series

$$\mathbf{u}(\mathbf{r},t) = \mathbf{U}(\theta,\mathbf{X},T) + \sum_{n=1}^{J} \mu^n \mathbf{u}^{(n)}(\theta,\mathbf{X},T), \tag{3.4}$$

where J is an integer determining the accuracy of the solution, and in the phase θ the parameters $\omega = \theta_t$ and $\mathbf{k} = -\nabla\theta$ can depend on slow variables $T = \mu t$ and $\mathbf{X} = \mu\mathbf{r}$.

In general, perturbations $\mathbf{u}^{(n)}$ may arbitrarily depend on $\mathbf{r}$ and t, depending on the initial conditions. As in the previous chapter, we concentrate on preventing resonances between perturbations and the main solution, which are possible when the perturbations depend on the same variable θ. In the nonlinear case, even at $\mu = 0$ the basic solution $\mathbf{U}$ can itself be unstable with respect to other types of small perturbations. Here we postulate the absence of such instabilities, so that all "free" solutions of the linearized equations remain finite. This justifies the representation (3.4) in which only the "forced" perturbations with a fast dependence only on θ are considered. As soon as an asymptotic solution is constructed, the non-steady perturbations can readily be constructed.

Now substitute the series (3.4) into (3.1) and expand the matrices A, $\mathbf{B}$, and C in a series of μ, assuming that they are sufficiently smooth. After zeroing the expressions at each power of μ, we obtain, in each order of approximation, a set of linear equations for perturbations:

$$G\mathbf{u}^{(n)} = \mathbf{H}^{(n)}. \tag{3.5}$$

Here, G is a linear operator which appears as a result of the linearization of Eqs (3.3) near the solution $\mathbf{U}(\theta, \mathbf{A})$, namely,

$$G\mathbf{u}^{(n)} = \left(\omega A^{(0)} - \mathbf{k}\mathbf{B}^{(0)}\right)\frac{\partial \mathbf{u}^{(1)}}{\partial\theta} + \left[\frac{\partial Q^{(0)}}{\partial \mathbf{U}} + \left(\omega\frac{\partial A^{(0)}}{\partial \mathbf{U}} - \mathbf{k}\frac{\partial \mathbf{B}^{(0)}}{\partial \mathbf{U}}\right)\frac{\partial \mathbf{U}}{\partial\theta}\right]\mathbf{u}^{(n)}. \tag{3.6}$$

The right-hand side of Eqs (3.5) depends only on the solutions obtained in previous approximations. In particular,

$$\mathbf{H}^{(1)} = \mathbf{f}^{(0)} - \left(A^{(0)} \frac{\partial \mathbf{U}}{\partial T} + \mathbf{B}^{(0)} \frac{\partial \mathbf{U}}{\partial \mathbf{X}} \right), \ \mathbf{f}^{(0)} = \mathbf{f}(\mathbf{U}, T, \mathbf{X}). \tag{3.7}$$

Since the matrices $A^{(0)}$, $\mathbf{B}^{(0)}$, and $C^{(0)}$ are now functions of $\mathbf{U}$, the linear equations (3.5) have variable coefficients which are, by definition, periodic in θ.

The theory of linear ODEs with periodic coefficients is well established and can be found in numerous mathematical textbooks [1, 11]. The general solution of (3.5) can be written in the form

$$\mathbf{u}^{(n)} = Y\left[\mathbf{C}^{(n)} + \int_0^{\theta} Y^{*}\mathbf{H}^{(n)} d\theta' \right]. \tag{3.8}$$

Here Y is the fundamental matrix composed of linearly independent solutions (vectors) $\mathbf{Y}_i$ of the homogeneous system $G\mathbf{u}^{(n)} = 0$; Y^* is its adjoint counterpart, so that

$$Y^{*} = Y^{-1}\left(\omega A^{(0)} - \mathbf{k}\mathbf{B}^{(0)} \right)^{-1}. \tag{3.9}$$

We recall that Y^{-1} is the inverted matrix, so that $YY^{-1} = I$, where I is the identity (unit) matrix. The vector $\mathbf{C}^{(n)}(T, \mathbf{X})$ is integration "constant" independent of the "fast" variable θ; it can, however, slowly vary in time and space.

It is important that as long as $\mathbf{U}$ is known, $m + 1$ vectors (columns) of the matrix Y are known as well. Indeed, after differentiating the basic system (3.3) with respect to θ and each A_i, one can see that the following vectors satisfy the linear homogeneous system $G\mathbf{u}^{(n)} = 0$:

$$\begin{aligned} &\mathbf{Y}_1 = \mathbf{U}_{\theta}, \quad \mathbf{Y}_i = \mathbf{U}_{A_i}, \quad i = 2,\ 3,\ \ldots\ m, \\ &\mathbf{Y}_{m+1} = \mathbf{Y}'_{m+1} + \alpha\theta\mathbf{U}_{\theta}\ . \end{aligned} \tag{3.10}$$

The vectors Y_1, $Y_2\ldots$, Y_m, and Y'_{m+1} are periodic functions of θ and α does not depend on θ. The secular term in Y_{m+1} appears from differentiation of (3.3) with respect to the parameters ω and **k** which are present in $\theta = \omega t - \mathbf{kr}$. Indeed, differentiation of **U** with respect to ω and **k** produces terms proportional to $t\mathbf{U}_\theta$ and $\mathbf{rU}_\theta$, respectively. Multiplying them respectively by ω and **k** and summing, we obtain a secular term proportional to $\theta\mathbf{U}_\theta$ which is also a solution of $G\mathbf{u}^{(n)} = 0$. Note that the matrix Y_{m+1} appears only in the nonlinear case when the wave period and length (or the wave velocity) depend on the parameters A_i such as wave amplitude.

The remaining $N - m$ vectors of the fundamental matrix satisfy linear equations with periodic coefficients and can be presented by means of the Floquet theory [5, 11]. Namely, $\mathbf{Y}_i \propto \exp(\lambda_i\theta)\psi_i(\theta)$, where $\psi_i(\theta)$ are periodic functions and λ_i are constants (characteristic exponents). It is assumed that if $\mathrm{Re}\,\lambda_i = 0$, then $\mathrm{Im}\,\lambda_i \neq 0, \pm 1, \pm 2\ldots$. The latter condition is known in the theory of nonlinear oscillations as the absence of internal resonances [6]. When this condition is met, the corresponding $N - m$ vectors $\mathbf{Y}_i$ are out of resonance with the fundamental periodic solution.

Thus, all elements of the resulting matrix Y can be written in the form

$$\begin{aligned} Y_{ij} &= e^{\lambda_j\theta} y_{ij} + \alpha_i \theta y_{i1} \delta_{j,m+1} + c.c., \\ Y^*_{jk} &= e^{-\lambda_j\theta} y^*_{jk} - \alpha_i \theta y^*_{m+1} \delta_{j,1} + c.c. \end{aligned} \tag{3.11}$$

Here, y_{ij} and y_{jk} are periodic functions of θ and $c.c.$ denotes complex conjugation. Substituting these matrices into (3.8), we obtain

$$u_i^{(n)} = \sum_{j,k} \left\{ \begin{aligned} & e^{\lambda_i\theta} y_{ij} C_j^{(n)} + \alpha\theta y_{i1} C_{m+1}^{(n)} \delta_{j,m+1} + e^{\lambda_j\theta} y_{ij} \int_0^\theta e^{-\lambda_j\theta} y^*_{jk} H_k^{(n)} d\theta' \\ & + \alpha\theta y_{i1} \int_0^\theta y^*_{m+1,j} H_j^{(n)} d\theta' \end{aligned} \right\} + c.c., \tag{3.12}$$

where y^*_{jk} are periodic functions of θ corresponding to the adjoint matrix Y^*. By virtue of (3.10), $\lambda_1 = ... = \lambda_{m+1} = 0$; in particular, $\mathbf{y}_1 = \mathbf{U}_\theta$. Here, the terms corresponding to the secular term in $\mathbf{Y}_{m+1}$ are written separately.

In general, the perturbation (3.12) diverges at $|\theta| \to \infty$ which can eventually destroy the validity of the solution. To prevent the divergence, the following constraints should be imposed. First, at j = 2,...m the integral in the upper line increases secularly unless

$$\int_0^{2\pi} y^*_{jk} H_k^{(n)} d\theta = 0, \qquad j = 2,...m, \tag{3.13}$$

so that the integral does not build up from period to period. The remaining secularities can be removed by choosing the constants $C^{(n)}$, namely,

$$2\pi C^{(n)}_{m+1} + \int_0^{2\pi} y^*_{1k} H_k^{(n)} d\theta + \alpha \int_0^{2\pi} d\theta \int_0^{\theta} y^*_{m+1,j} H_j^{(n)} d\theta' = 0, \tag{3.14}$$

and finally, for $\lambda \neq 0$,

$$C_j^{(n)} = \begin{cases} -\int_0^{\infty} y^*_{ik} H_k^{(n)} e^{-\lambda_j \theta} d\theta & \text{if } \mathrm{Re}\lambda_j > 0, \\ \int_{-\infty}^{0} y^*_{ik} H_k^{(n)} e^{-\lambda_j \theta} d\theta & \text{if } \mathrm{Re}\lambda_j < 0. \end{cases} \tag{3.15}$$

At $\mathrm{Re}\lambda_j = 0$ the aforementioned condition of the absence of internal resonances is necessary for the finiteness of the perturbations.

Considering the form of $\mathbf{H}^{(1)}$ as given by (3.7) and similarly for higher-order approximations, one can see that compatibility (orthogonality) conditions (3.13) represent m – 1 differential equations for slow variation of the components of vector $\mathbf{A}$ as functions of T and $\mathbf{X}$.

Note, however, that there is no independent equation defining perturbations of the wave phase or frequency; as known from the oscillation theory [6], they appear in the next approximation. To secure a transition to the linear case, we use this ambiguity to add one more orthogonality condition:

$$\int_0^{2\pi} y_{1k}^* H_k^{(n)} d\theta = 0. \tag{3.16}$$

In the linear case, when ω is independent of $\mathbf{A}$ and $\alpha = 0$, this condition becomes necessary in the first approximation.

In the integral expressions one should regard $\mathbf{A}^{(n)}$, ω, and $\mathbf{k}$ as unknown, slowly varying functions. As in the previous chapter, it is necessary to add the obvious "kinematic" equations for $\omega = \theta_t$ and $\mathbf{k} = -\nabla\theta$:

$$\frac{\partial \mathbf{k}}{\partial T} + \nabla \cdot \omega = 0, \quad \nabla \times \mathbf{k} = 0. \tag{3.17}$$

This completes the construction of a closed set of equations determining slow variation of the wave parameters.

As already mentioned, in the original papers [3, 7] the above method was described for an even more general system of equations:

$$\mathbf{M}\left(\frac{\partial \mathbf{u}}{\partial t}, \frac{\partial \mathbf{u}}{\partial \mathbf{r}}, \mu\right) = 0, \tag{3.18}$$

where now $\mathbf{M}$ is a sufficiently smooth but not necessarily quasi-linear vector, and μ is, as above, a small parameter.

Assume that at $\mu = 0$ the system (3.18) has a periodic solution in the form of a plane wave (3.2). This solution satisfies the unperturbed set of ODEs,

$$\mathbf{M}^{(0)} = \mathbf{M}(\omega\mathbf{U}_\theta, -\mathbf{k}\mathbf{U}_\theta, \mathbf{U}) = 0. \tag{3.19}$$

For a perturbed system, when $\mu \neq 0$, the solution is represented by a series (3.4), to result in the linear system (3.5), where now

$$G = \left(\omega \frac{\partial \mathbf{M}^{(0)}}{\partial \mathbf{U}_t} - \mathbf{k} \frac{\partial \mathbf{M}^{(0)}}{\partial \mathbf{U}_\mathbf{r}} \right) \frac{d}{d\theta} + \frac{\partial \mathbf{M}^{(0)}}{\partial \mathbf{U}}, \tag{3.20}$$

and

$$\mathbf{H}^{(1)} = \frac{\partial \mathbf{M}^{(0)}}{\partial \mu} - \frac{\partial \mathbf{M}^{(0)}}{\partial \mathbf{U}_t} \mathbf{U}_T - \frac{\partial \mathbf{M}^{(0)}}{\partial \mathbf{U}_\mathbf{r}} \mathbf{U}_\mathbf{X}. \tag{3.21}$$

Here $\partial / \partial\mu$ is understood as a derivative with respect to the unperturbed parameters of the order of μ; in particular, in (3.7) it is $\partial \mathbf{M}^{(0)} / \partial\mu = \mathbf{f}^{(0)}$.

3.2. Lagrangian Systems

The above perturbation scheme allows a more vivid physical interpretation if the basic equations are written in the Lagrangian form. In this section we consider the Euler–Lagrange equations which follow from the Hamilton's variation principle.

It is assumed that the Lagrangian density $L(\mathbf{u}, \mathbf{u}_t, \nabla\mathbf{u})$ is known for the unperturbed system; as it is generally accepted, we shall call L simply the Lagrangian. From the classical principle of least action:

$$\delta S = \iint_D \left(\frac{\partial L}{\partial \mathbf{u}} \delta\mathbf{u} + \frac{\partial L}{\partial \mathbf{u}_t} \delta\mathbf{u}_t + \frac{\partial L}{\partial \mathbf{u}_\mathbf{r}} \delta\mathbf{u}_\mathbf{r} \right) dt d\mathbf{r} = 0, \tag{3.22}$$

where $\delta\mathbf{u}$, $\delta\mathbf{u}_t$, and $\delta\mathbf{u}_\mathbf{r}$ are variations of the field $\mathbf{u}(t, \mathbf{r})$ and its derivatives with respect to the real motion, and D is the domain where the wave field is determined (in our case D is fixed), the Euler–Lagrange equations follow [2, 4]:

$$\frac{\partial}{\partial t}\frac{\partial L}{\partial \mathbf{u}_t}+\frac{\partial}{\partial \mathbf{r}}\frac{\partial L}{\partial \mathbf{u}_\mathbf{r}}-\frac{\partial L}{\partial \mathbf{u}}=0. \tag{3.23}$$

Let us consider here a perturbed system based on (3.23) in which the Lagrangian can contain slowly varying parameters, i.e., $L = L(\mathbf{u}, \mathbf{u}_t, \mathbf{u}_\mathbf{r}, T, \mathbf{X})$:

$$\frac{\partial}{\partial t}\frac{\partial L}{\partial \mathbf{u}_t}+\frac{\partial}{\partial \mathbf{r}}\frac{\partial L}{\partial \mathbf{u}_\mathbf{r}}-\frac{\partial L}{\partial \mathbf{u}}=\mu\mathbf{R}. \tag{3.24}$$

Here, as before, $\mathbf{u}$ is an N-dimensional vector of unknown functions, and the function $\mathbf{R}(\mathbf{u},\mathbf{u}_t,\mathbf{u}_\mathbf{r},T,\mathbf{X})$ is responsible for small perturbing factors such as dissipation. We again assume that at $\mu=0$ this system has a family of particular solutions in the form of a plane periodic progressive wave, $\mathbf{u} = \mathbf{U}(\theta,\mathbf{A})$, $\theta=\omega t-\mathbf{kr}+\theta_0$, satisfying the set of ODEs:

$$\frac{\partial}{\partial \theta}\left(\omega\frac{\partial L}{\partial \mathbf{u}_t}-\mathbf{k}\frac{\partial L}{\partial \mathbf{u}_\mathbf{r}}\right)-\frac{\partial L}{\partial \mathbf{u}}=0. \tag{3.25}$$

Representing the solution as a series (3.4):

$$\mathbf{u}(\mathbf{r},t)=\mathbf{U}(\theta,T,\mathbf{X})+\sum_{n=1}^{J}\mu^n\mathbf{u}^{(n)}(\theta,T,\mathbf{X}) \tag{3.26}$$

and substituting it into (3.24), we obtain the corresponding expansions in each order. In particular, up to the first order of μ we have

$$\frac{\partial L}{\partial \mathbf{u}} = \frac{\partial L}{\partial \mathbf{U}} + \mu\left\{\frac{\partial^2 L}{\partial \mathbf{U}^2}\mathbf{u}^{(1)} + \frac{\partial^2 L}{\partial \mathbf{U}\partial \mathbf{U}_t}\left(\mathbf{u}_t^{(1)} + \mathbf{U}_T\right) + \frac{\partial^2 L}{\partial \mathbf{U}\partial \mathbf{U}_{\mathbf{r}}}\left(\mathbf{u}_{\mathbf{r}}^{(1)} + \mathbf{U}_{\mathbf{X}}\right)\right\},$$

$$\frac{\partial L}{\partial \mathbf{u}_t} = \frac{\partial L}{\partial \mathbf{U}_t} + \mu\left\{\frac{\partial^2 L}{\partial \mathbf{U}\partial \mathbf{U}_t}\mathbf{u}^{(1)} + \frac{\partial^2 L}{\partial \mathbf{U}_t^2}\left(\mathbf{u}_t^{(1)} + \mathbf{U}_T\right) + \frac{\partial^2 L}{\partial \mathbf{U}\partial \mathbf{U}_{\mathbf{r}}}\left(\mathbf{u}_{\mathbf{r}}^{(1)} + \mathbf{U}_{\mathbf{X}}\right)\right\},$$

$$\frac{\partial}{\partial t}\frac{\partial L}{\partial \mathbf{u}_t} = \frac{\partial}{\partial t}\frac{\partial L}{\partial \mathbf{U}_t} + \mu\left\{\frac{\partial}{\partial T}\frac{\partial L}{\partial \mathbf{U}_t} + \omega\frac{\partial}{\partial \theta}\left[\begin{array}{l}\dfrac{\partial^2 L}{\partial \mathbf{U}\partial \mathbf{U}_t}\mathbf{u}^{(1)} + \dfrac{\partial^2 L}{\partial \mathbf{U}_t^2}\left(\mathbf{u}_t^{(1)} + \mathbf{U}_T\right)\\ + \dfrac{\partial^2 L}{\partial \mathbf{U}\partial \mathbf{U}_{\mathbf{r}}}\left(\mathbf{u}_{\mathbf{r}}^{(1)} + \mathbf{U}_{\mathbf{X}}\right)\end{array}\right]\right\}, \tag{3.27}$$

$$\mathbf{R} = \mathbf{R}^{(0)}.$$

Here the terms of type $(\partial^2 L/\partial \mathbf{U}^2)\mathbf{u}^{(1)}$ denote multiplication of the matrix $\partial^2 L/\partial \mathbf{U}^2$ by the column vector $\mathbf{u}^{(1)}$. Analogous expansions can be written for the functions $\partial L/\partial \mathbf{u}_{\mathbf{r}}$ and $\partial\,(\partial L/\partial \mathbf{u}_{\mathbf{r}})/\,\partial \mathbf{r}$. The terms of the order of μ^2 and higher are neglected. In the right-hand sides of these equations the Lagrangian is taken as $L = L(\mathbf{U}, \omega\mathbf{U}_\theta, -k\mathbf{U}_\theta, T, \mathbf{X})$. Substituting (3.26) and (3.27) into (3.24) and taking into account (3.25), we obtain, similarly to (3.5), a linear system

$$G_L \mathbf{u}^{(n)} = \mathbf{H}^{(n)}, \tag{3.28}$$

where

$$G_L \mathbf{u}^{(n)} = \frac{d}{d\theta}\left(\frac{\partial^2 L}{\partial \mathbf{U}_\theta^2}\mathbf{u}_\theta^{(n)} + \frac{\partial^2 L}{\partial \mathbf{U}\partial \mathbf{U}_\theta}\mathbf{u}^{(n)}\right) - \frac{\partial^2 L}{\partial \mathbf{U}_\theta^2}\mathbf{u}^{(n)} - \frac{\partial^2 L}{\partial \mathbf{U}\partial \mathbf{U}_\theta}\mathbf{u}_\theta^{(n)}, \tag{3.29}$$

and, in the first approximation,

$$\mathbf{H}^{(1)} = \mathbf{R}^{(0)} + \frac{\partial^2 L}{\partial \mathbf{U}\partial \mathbf{U}_t}\mathbf{U}_T + \frac{\partial^2 L}{\partial \mathbf{U}\partial \mathbf{U}_{\mathbf{r}}}\mathbf{U}_{\mathbf{X}} - \frac{\partial}{\partial T}\frac{\partial L}{\partial \mathbf{U}_t} - \frac{\partial}{\partial \mathbf{X}}\frac{\partial L}{\partial \mathbf{U}_{\mathbf{r}}} -$$
$$-\omega\frac{d}{d\theta}\left(\frac{\partial^2 L}{\partial \mathbf{U}_t^2}\mathbf{U}_T + \frac{\partial^2 L}{\partial \mathbf{U}_t\partial \mathbf{U}_{\mathbf{r}}}\mathbf{U}_{\mathbf{X}}\right) + \mathbf{k}\frac{d}{d\theta}\left(\frac{\partial^2 L}{\partial \mathbf{U}_{\mathbf{r}}\partial \mathbf{U}_t}\mathbf{U}_T + \frac{\partial^2 L}{(\partial \mathbf{U}_{\mathbf{r}})^2}\mathbf{U}_{\mathbf{X}}\right). \tag{3.30}$$

Thus, we again have an ODE set differing from Eq. (3.5) in that the operator G_L is of the second order. Still the general solution of (3.38) can be represented in the form (3.8):

$$\mathbf{u}^{(n)} = Y\left[\mathbf{C}^{(n)} + \int_0^\theta Y^*\mathbf{H}^{(n)} d\theta'\right], \tag{3.31}$$

where the fundamental matrix Y of solutions of the equations $G_L = 0$ is not square but it has $N \times 2N$ dimension, so that the matrix Y^* is $2N \times N$.

In this case at least two particular solutions are known *a priori*:

$$\mathbf{Y}_1 = \mathbf{U}_\theta, \quad \mathbf{Y}_2 = \mathbf{Y}_2' + \alpha\theta\mathbf{U}_\theta. \tag{3.32}$$

The remaining $2N$–2 vectors again have the form prescribed by the Floquet theorem, so that the matrices Y and Y^* have the form (3.11) with m = 1. The general solution (3.31) can be written in the form similar to (3.12):

$$\begin{aligned} u_i^{(n)} &= y_{ij}C_j^{(n)} \exp(\lambda_j\theta) + a\theta y_{i1}C_2^{(n)} + \\ & y_{ij}\exp(\lambda_j\theta)\int_0^\theta y_{jk}^* H_k^{(n)} \exp(-\lambda_j\theta')d\theta' \\ & + a y_{i1}\int_0^\theta d\theta' \int_0^{\theta'} y_{2k}^* H_k^{(n)} d\theta'' + \text{c.c.} \end{aligned} \tag{3.33}$$

Here, the column y_{i1} is the same as $\mathbf{U}_\theta$. The conditions for non-increasing perturbations are the same as (3.13) to (3.16) with m = 1. An important new factor here as compared to the previous section is that the operator G_L in (3.29) is self-adjoint. In other words, its matrix is Hermitian, i.e., it remains unchanged after being transposed (rows are changed to columns) and complex conjugated. Here, it can be readily verified by representing G_L in the standard form $G_L = \sum s_k(\theta)d^k u / d\theta^n$, so that its formal adjoint matrix is $G_L^* = \sum(-1)^k d^k [s_k(\theta)u] / d\theta^n$, that $G_L = G_L^*$. In this case the conditions (3.16) and (3.13) can be written as

$$\int_0^{2\pi} \mathbf{U}_\theta H_k^{(n)} d\theta = 0, \qquad \int_0^{2\pi} \mathbf{U}_\mathbf{A} H_k^{(n)} d\theta = 0. \tag{3.34}$$

3.3. Averaged Lagrangian and Whitham's Variational Principle

As shown in [8], if $\mathbf{R}^{(0)} = 0$, then in the first approximation the orthogonality conditions (3.34) allow the Lagrangian formulation of the equations for slowly varying parameters of the solution. For this purpose we substitute (3.30) into the first equation (3.34):

$$\begin{aligned} \int_0^{2\pi} \mathbf{U}_\theta \Bigg[& \omega \frac{d}{d\theta}\left(\frac{\partial^2 L}{\partial \mathbf{U}_t^2} \mathbf{U}_T + \frac{\partial^2 L}{\partial \mathbf{U}_t \partial \mathbf{U}_\mathbf{r}} \mathbf{U}_\mathbf{X} \right) \\ & -\mathbf{k} \frac{d}{d\theta}\left(\frac{\partial^2 L}{\partial \mathbf{U}_\mathbf{r} \partial \mathbf{U}_t} \mathbf{U}_T + \frac{\partial^2 L}{(\partial \mathbf{U}_\mathbf{r})^2} \mathbf{U}_\mathbf{X} \right) - \frac{\partial^2 L}{\partial \mathbf{U} \partial \mathbf{U}_t} \mathbf{U}_T \\ & - \frac{\partial^2 L}{\partial \mathbf{U} \partial \mathbf{U}_\mathbf{r}} \mathbf{U}_\mathbf{X} + \frac{\partial}{\partial T} \frac{\partial L}{\partial \mathbf{U}_t} + \frac{\partial}{\partial \mathbf{X}} \frac{\partial L}{\partial \mathbf{U}_\mathbf{r}} \Bigg] d\theta = 0. \end{aligned} \tag{3.35}$$

Now integrate this by parts, taking into account the periodicity with respect to θ. After simple transformations, (3.35) can be written in the form

$$\int_0^{2\pi} \left[\frac{\partial}{\partial T}\left(\frac{\partial L}{\partial \mathbf{U}_t} \mathbf{U}_\theta \right) + \frac{\partial}{\partial \mathbf{X}} \frac{\partial L}{\partial \mathbf{U}_\mathbf{r}} \mathbf{U}_\theta \right] d\theta = 0. \tag{3.36}$$

The same result follows from variation of the period-averaged Lagrangian

$$\mathcal{L}(\omega,\mathbf{k},\mathbf{A})=\frac{1}{2\pi}\int_0^{2\pi} L d\theta \tag{3.37}$$

with respect to the variable θ represented by its slowly varying derivatives, $\omega=\theta_t$ and $\mathbf{k}=-\nabla\theta$. In other words, if $\mathbf{R}^{(0)}=0$, we have

$$\frac{\partial}{\partial T}\frac{\partial\mathcal{L}}{\partial\omega}-\frac{\partial}{\partial\mathbf{X}}\frac{\partial\mathcal{L}}{\partial\mathbf{k}}=0. \tag{3.38}$$

Here the functions are understood as derivatives with respect to the explicitly present ω and k:

$$\frac{\partial\mathcal{L}}{\partial\omega}=\frac{1}{2\pi}\int_0^{2\pi}\mathbf{U}_\theta\frac{\partial L}{\partial\mathbf{U}_t}d\theta,\quad \frac{\partial\mathcal{L}}{\partial\mathbf{k}}=\frac{1}{2\pi}\int_0^{2\pi}\mathbf{U}_\theta\frac{\partial L}{\partial\mathbf{U}_\mathbf{r}}d\theta. \tag{3.39}$$

The function $J=\partial\mathcal{L}/\partial\omega$ is the average wave action density. As follows from (3.38), for any localized wave packet, the total action, $I=\int_{-\infty}^{\infty}Jdx$, is conserved.

Similarly, using the second equation (3.34), we obtain the nonlinear dispersion equation relating **A**, ω, and k:

$$\frac{\partial\mathcal{L}}{\partial\mathbf{A}}=0. \tag{3.40}$$

Equations (3.39) and (3.40) express the averaged variational principle suggested by Whitham [10]: variation of the period-averaged Lagrangian with respect to the wave phase and amplitude yields the equations for the slowly varying parameters of the wave. Thus, the above results can be considered the derivation of this principle as the first approximation of the perturbation theory for the Lagrangian systems.

Introducing period-averaged wave energy and energy flux,

$$E = \omega\frac{\partial \mathcal{L}}{\partial \omega} - \mathcal{L}, \quad \mathbf{S} = -\omega\frac{\partial \mathcal{L}}{\partial \mathbf{k}}, \tag{3.41}$$

we obtain the energy transfer equation

$$\frac{\partial E}{\partial T} + \frac{\partial \mathbf{S}}{\partial \mathbf{X}} = 0. \tag{3.42}$$

For the average wave momentum vector and the momentum flux tensor, we have

$$\mathbf{G} = -\mathbf{k}\frac{\partial \mathcal{L}}{\partial \omega}, \quad \Pi_{ik} = -k_i\frac{\partial \mathcal{L}}{\partial k_k} + \mathcal{L}\delta_{ik}, \tag{3.43}$$

with the corresponding transfer equation

$$\frac{\partial G_i}{\partial T} + \frac{\partial \Pi_{ik}}{\partial X_k} = \frac{\partial \mathcal{L}}{\partial X_i}. \tag{3.44}$$

Now we return to the full equation (3.24) with $\mathbf{R}^{(0)} \neq 0$. Using an analogy with classical mechanics, we introduce the density of the dissipative Rayleigh function Q such as $\mathbf{R} = -\partial Q/\partial \mathbf{U}_t$. Denoting the period-averaged function Q by Q_R, we generalize (3.38) as

$$\frac{\partial}{\partial T}\frac{\partial \mathcal{L}}{\partial \omega} - \frac{\partial}{\partial \mathbf{X}}\frac{\partial \mathcal{L}}{\partial \mathbf{k}} = -\mu\frac{\partial Q_R}{\partial \omega}. \tag{3.45}$$

As shown in [8], Eq. (3.45) is valid in a more general case when the term with $\mathbf{R}$ in (3.24) is not small, and there is no μ in the right-hand part of (3.45).

As above, the kinematic equations (3.17) should be added here to close the set of equations.

Now one can see some physics behind these transformations.

3.4. Linear Waves

The Lagrangian approach gives especially simple results for quasi-harmonic waves in linear equations. In this case the averaged Lagrangian (as well as the wave energy) is a quadratic function of the wave amplitude. Moreover, in the linear case, using Eqs (2.4) from the previous chapter, amplitudes of all components of the vector **u** can be expressed by means of one of them, which we denote a:

$$\mathcal{L} = f(\omega, \mathbf{k}, T, \mathbf{X}) a^2 (T, \mathbf{X}). \tag{3.46}$$

Substituting this into (3.40), we obtain the dispersion equation, relating ω and k at each T and $\mathbf{X}$:

$$f(\omega, \mathbf{k}, T, \mathbf{X}) = 0. \tag{3.47}$$

If this equation can be resolved as $\omega = \omega(\mathbf{k}, T, \mathbf{X})$, substituting this into the first equation (3.17) yields

$$\frac{\partial \mathbf{k}}{\partial T} + \left(\mathbf{c}_g \cdot \nabla_{\mathbf{X}}\right)\mathbf{k} = -\nabla_{\mathbf{X}} \cdot \omega. \tag{3.48}$$

Here $\mathbf{c}_g = \partial\omega / \partial\mathbf{k}$ is the group velocity vector, and $\nabla_{\mathbf{X}}$ denotes differentiation with respect to slow coordinates.

As is evident from (3.46) and (3.47), in the linear case $\mathcal{L} = 0$. Then from (3.41) it is easy to see that, as expected, $\mathbf{S} = \mathbf{c}_g E$. Besides, in many cases, $\mathcal{L} = K - P$, where K and P are, respectively, the kinetic and potential energy densities, so that in our case these energies are, on average, equal. The latter is a wave analog of the virial theorem known in classical mechanics.

The wave amplitude can be found from Eqs (3.44) or (3.45). If Q_R = 0, it is convenient to use the energy transfer equation (3.42) which yields

$$\frac{\partial E}{\partial T}+\frac{\partial(\mathbf{c}_g E)}{\partial \mathbf{X}}=0. \tag{3.49}$$

In view of (3.41) and (3.46), this can be considered as an equation for the wave amplitude.

In a medium with constant parameters, when in (3.46) f does not explicitly depend on T and X, the above equations can be solved analytically [9]. In the one-dimensional case when X and c_g are scalars, (3.48) reduces to the nonlinear (due to the dependence $c_g(k)$) equation

$$\frac{\partial k}{\partial T}+c_g\frac{\partial k}{\partial X}=0, \tag{3.50}$$

with a solution in the form of a *simple wave*:

$$k=\kappa[X-c_g T], \text{ or } X-c_g T=\vartheta(k), \tag{3.51}$$

where κ and ϑ are arbitrary functions which should be defined by the initial condition. We shall discuss properties of such simple "waves of envelopes" later in Chapters 4 and 7; now we use (3.51) to find the solution of the energy equation (3.49) in the one-dimensional case. Since c_g can be considered as a function of $k(\vartheta)$, we pass from T and X to the variables $T' = T$ and ϑ. Differentiating ϑ and using (3.51), it is easy to see that

$$\frac{\partial \vartheta}{\partial X}=-\frac{1}{c_g}\frac{\partial \vartheta}{\partial X}=\frac{1}{1+T'dc_g/d\vartheta}. \tag{3.52}$$

Substituting this into (3.49), we obtain a separable equation

$$\frac{\partial E}{\partial T'}+\frac{E dc_g/d\vartheta}{1+T'dc_g/d\vartheta}=0, \tag{3.53}$$

with a solution

$$E=\frac{W(\vartheta)}{1+T'dc_g/d\vartheta}. \tag{3.54}$$

Here W is an arbitrary function. Since E is proportional to the square of amplitude, the latter has a similar form

$$a = \frac{s(\vartheta)}{\sqrt{1 + T' dc_g / d\vartheta}}. \tag{3.55}$$

The properties of these solutions (including "dispersion compression" when E and a diverge at some point) are considered in [9]. We shall use these solutions below in this book.

3.5. Concluding Remarks

The asymptotic perturbation scheme discussed in this chapter can look cumbersome but its main idea is the same as above: first, linear equations are written for perturbations and second, the orthogonality conditions prevent the growth of these perturbations due to their resonances with the perturbing factor. These conditions lead to the equations for slowly varying parameters of the solutions obtained in previous approximations, most importantly in zero approximation. These schemes are not limited by integrability of the basic equations. In the Lagrangian case, the final results in the first approximation have a rather simple form, corresponding to the averaged variational principle and its generalizations.

In the next chapter we consider some realistic examples of nonlinear waves.

References

1. Gantmacher, F. R. (2000). Chapter XIV in *Matrix Theory*, v. II. AMS Chelsea, Providence, Rhode Island.
2. Gelfand, I. M. and Fomin, S. V. (1963). *Calculus of Variations*. Dover, New York.
3. Gorshkov, K. A., Ostrovsky, L. A., and Pelinovsky, E. N. (1974). Some problems of asymptotic theory of nonlinear-waves, *Proceedings of the Institute of Electrical and Electronics Engineers*, v. 62, pp. 1511–1517.
4. Landau, L. D. and Lifshitz, E. M. (1975). *The Classical Theory of Fields*, v. 2. 4th Ed. Butterworth-Heinemann, Oxford.
5. McLachlan, N. W. (1964). *Theory and Application of Mathieu Function.* Dover, New York.

6. Mitropol'skii, Y. A. (1965). *Problems of the asymptotic theory of nonstationary vibrations*. Israel Program for Scientific Translations, Jerusalem.
7. Ostrovsky, L. A. and Pelinovsky, E. N. (1971). Averaging method for non-sinusoidal waves, *Soviet Physics – Doklady*, v. 15, pp. 1097–1099.
8. Ostrovsky, L. A. and Pelinovsky, E. N. (1972). Averaging method and generalized variation principle for nonsinusoidal waves, *Applied Mathematics and Mechanics*, v. 36, pp. 63–70.
9. Ostrovsky, L. A. and Potapov, A. I. (1999). *Modulated Waves: Theory and Applications.* Johns Hopkins University Press, Baltimore.
10. Whitham, G. B. (1974). *Linear and Nonlinear Waves.* J. Wiley & Sons, New York.
11. Yakubovich, V. A. and Starzhinskii, V. M. (1975). Chapter II in *Linear differential equation with periodic coefficients.* J. Wiley & Sons, New York and Toronto.

Chapter 4

Nonlinear Waves of Modulation

> The aim of mathematics is to explain as much as possible in simple terms.
>
> Sir Michael Atiyah

4.1. Simple Envelope Waves

Let us now consider some particular cases of "waves of modulation" based on the method described in the previous chapter. In the one-dimensional case when a plane wave does not change its propagation direction, Eqs (3.17) and (3.16) or (3.38) of Chapter 3 can be represented in the form

$$\begin{aligned} &\frac{\partial k}{\partial T}+\frac{\partial \omega(k,A)}{\partial X}=0, \\ &\frac{\partial F_1(k,A)}{\partial T}+\frac{\partial F_2(k,A)}{\partial X}=0. \end{aligned} \tag{4.1}$$

Here $F_{1,2}$ are slowly varying functions, for example, average energy density and energy flux along the x-axis as in (3.42); these functions are known as long as the family of stationary solutions, $U(\theta,\mathbf{A})$, is known. A is a wave parameter, such as amplitude (for simplicity, here we consider only one such parameter), and $\omega(k,A)$ is determined by the nonlinear dispersion equation, also following from the stationary solution. The type of this second-order system can be seen from its particular solutions in which all variables are the functions of only one of them, for example, $k = k(A)$. Then from (4.1) it follows that

$$\begin{aligned} &k_A \frac{\partial A}{\partial T} + (\omega_A + \omega_k k_A)\frac{\partial A}{\partial X} = 0, \\ &(F_{1k}k_A + F_{1A})\frac{\partial A}{\partial T} + (F_{2k}k_A + F_{2A})\frac{\partial A}{\partial X} = 0. \end{aligned} \tag{4.2}$$

If all the functions are differentiable, these equations can be considered as a homogeneous algebraic system with respect to $\partial A / \partial T$ and $\partial A / \partial X$ which has a non-trivial solution only if its determinant is equal to zero. As a result, at any fixed A and k we have a quadratic equation for k_A:

$$\begin{aligned} &ak_A^2 + bk_A + c = 0, \\ &a = F_{2k} - \omega_k F_{1k}, b = F_{2A} - \omega_k F_{1A} - \omega_A F_{1k}, c = -\omega_A. \end{aligned} \tag{4.3}$$

At $b^2 - 4ac > 0$ we obtain two real values for k_A and then for each of them, real $k = k(A)$ as a solution of the corresponding first-order differential equation. Substituting this solution into any one of Eqs (4.2), we obtain the first-order nonlinear equation for A with the real propagation velocities:

$$\frac{\partial A}{\partial T} + V_s(A)\frac{\partial A}{\partial X} = 0, \ V_s(A) = \frac{\omega_A + \omega_k k_A}{k_A}. \tag{4.4}$$

where $V_s(A)$ can be considered as the nonlinear group velocity.

In this case, system (4.1) is hyperbolic; it is often called a "system of hydrodynamic type." Indeed, one-dimensional equations of the same structure are used in the classical fluid dynamics [1, 5] as well as in acoustics, magnetic hydrodynamics and other areas of physics, where they describe instantaneous physical values (e.g., pressure) rather than slowly varying parameters which can be (perhaps too broadly) called "envelopes," as already mentioned in Chapter 1. Their properties have been studied in detail. We remind readers that the solution of (4.4) for each function $U(A)$ has an implicit form:

$$A = f\left[X - V_s(A)T\right], \tag{4.5}$$

or

$$X - V_s(A)T = \Psi(A). \tag{4.6}$$

Here f and Ψ are arbitrary functions which should be defined by the initial condition.

This solution is a simple envelope wave similar to that obtained in Chapter 3 as the solution (3.51) for the slowly varying wave number. Each point of this wave's profile with a fixed A propagates at a constant velocity, or along a straight line, a *characteristic*, on the (X, T) plane. The parts of the wave with different levels of A move with different velocities, so that some parts of a wave can stretch ("rarefaction wave"), whereas other parts may steepen and eventually overturn, so that the wave profile becomes double-or triple-valued (Fig. 4.1). Near the overturning point Eq. (4.4) becomes insufficient, and some additional terms should be added, similarly to what was shown in Chapter 1 where Eq. (1.44) was changed to (1.46).

In the linear limit when $\omega_A = 0$, from (4.4) it follows that $V_s = c_g = \partial\omega / \partial k$ as expected. If nonlinearity is present but weak whereas dispersion is strong so that the basic wave $V_s(\theta, A)$ is close to a sinusoid, in many cases the velocity can be represented as $V_s = c_g(k) + q(k)a^2$, which leads to Eqs (1.44) and (1.46).

If, however, $b^2 - 4ac < 0$, the functions k_A and U are complex, which means that system (4.2) is elliptic. In particular, if a stationary, periodic wave train with constant $A = A_0$ and $k = k_0$ is slightly perturbed, substituting $A = A_0 + a_1 = A_0 + a\exp i(KX - \Omega T)$ into (4.2) or (4.4), we readily obtain $\Omega = V_0 K$, where $V_0 = V_s(A_0)$. Thus, at a real K and complex V_0, the modulation frequency Ω is complex. In general this means that the perturbation a_1 exponentially increases. This is the effect of self-modulation or modulation instability which was described in Chapter 1 for quasi-harmonic waves.

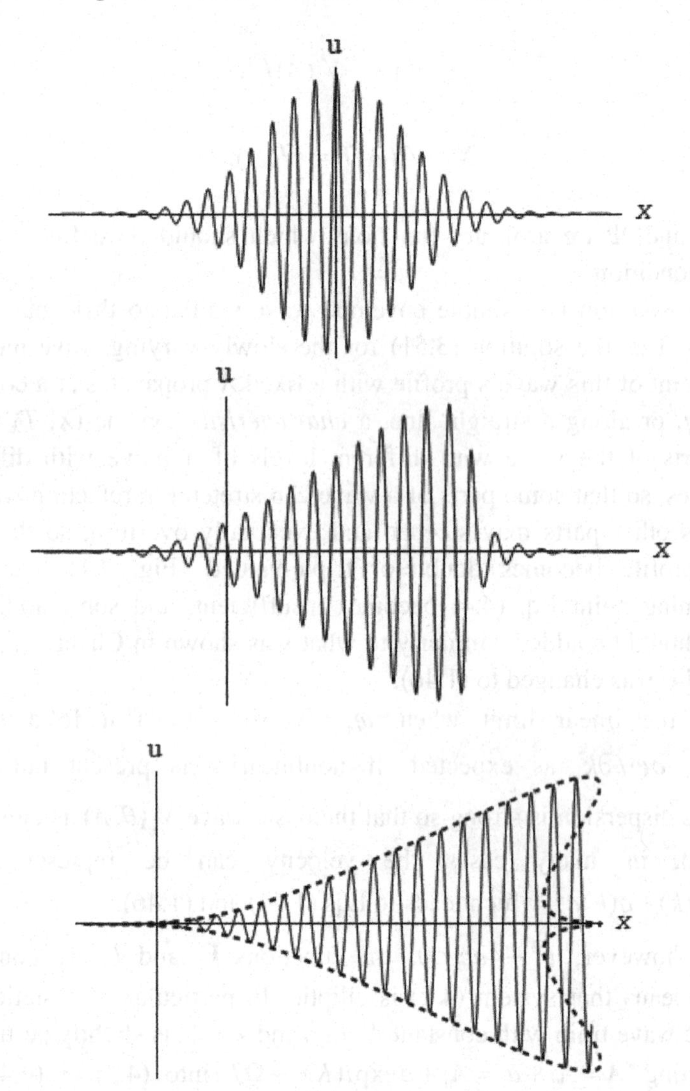

Figure 4.1. Schematic evolution of a wave train with an initially Gaussian envelope, in three subsequent time moments. Dashed line – the function (4.5) after overturning.

We do not further develop this important topic here because it has been discussed in many books and reviews, mainly for carrier waves close to harmonic, for which Eqs (1.44) or (1.46) can be used.

The particular case of a simple wave is the "rarefaction wave": at large times the wave train becomes much longer than the initial wave packet. At this stage of propagation, the initial impulse can be considered as delta-function, although physically our theory begins to work after the wave is expanded so that the scales of amplitude and period variations become large when compared to the period of carrier oscillations. At this stage the problem does not contain any separate parameter of length and time so that all slowly varying functions depend on a single variable, $\xi = X/T$ and thus, $\partial/\partial T = -(X/T^2)\partial/\partial\xi$, $\partial/\partial X = (1/T)\partial/\partial\xi$. Substituting this into (4.4), we have, at $A \neq const.$,

$$V_s(A) = \xi = X/T. \tag{4.7}$$

Hence, the asymptotic behavior of the amplitude in the "rarefaction wave" is such that the propagation velocity varies as X in space and $1/T$ in time. This is how the far left side ("tail") of the pulse shown in Fig. 4.1 behaves.

In general, there can be several possible velocities of type (4.7) and, correspondingly, several possible self-similar, "rarefaction" envelope waves.

4.2. Nonlinear Klein–Gordon Equation

4.2.1. *Averaged equations*

As an example of modulated non-harmonic waves, consider the so-called nonlinear Klein–Gordon equation [6, 10]:

$$u_{tt} - u_{xx} + P'(u) = 0. \tag{4.8}$$

Here P can be considered as potential and $P'(u)$ as the force acting in the wave system. At different functions $P(u)$ this equation describes a wide range of processes in, for example, crystals and superconductors, and was used in models of field theory. Here we shall obtain the averaged equations following Luke [6].

As above, suppose that (4.8) has solutions in the form of a periodic progressive wave, $u(x,t) = U(\theta = \omega t - kx)$, and seek the solution in the form

$$u(x,t) = U(\theta,T,X) + \mu u_1(\theta,T,X) + \mu^2 u_2(\theta,T,X) + \dots \ , \quad (4.9)$$

with slowly varying wave number and frequency defined as derivatives of θ:

$$\theta_t = \omega(T,X), \ \theta_x = -k(T,X) \quad (4.10)$$

As before,

$$k_T + \omega_X = 0. \quad (4.11)$$

Substituting (4.9) into (4.8), in the main and the first orders of μ we have

$$(\omega^2 - k^2)U_{\theta\theta} + P'(U) = 0 \quad (4.12)$$

and

$$\begin{aligned} G(u_1) &= (\omega^2 - k^2)u_{1\theta\theta} + P''(U)u_1 = H_1(\theta,T,X), \\ H_1 &= 2\omega U_{\theta T} + 2kU_{\theta X} + \omega_T U_\theta + k_X U_\theta . \end{aligned} \quad (4.13)$$

As above, the left-hand side of (4.13) is the variation of the basic equation (4.12), and, hence, $u_1 = U_\theta$ is a particular solution of the homogeneous equation $G(u_1) = 0$, i. e.,

$$(\omega^2 - k^2)U_{\theta\theta\theta} + P''(U)U_\theta = 0. \quad (4.14)$$

The same follows from differentiation of (4.12) with respect to θ. It is easy to verify that Eqs (4.12) and (4.14) are equivalent to

$$(\omega^2 - k^2)\partial(u_{1\theta}U_\theta - u_1 U_{\theta\theta}) / \partial\theta = U_\theta H_1, \tag{4.15}$$

or

$$u_{1\theta}U_\theta - u_1 U_{\theta\theta} = \frac{1}{\omega^2 - k^2}\int\int_0^\theta U_\theta H_1 d\theta'. \tag{4.16}$$

This equation is easily integrable. To keep u_1 bounded, it is necessary that the right-hand side of (4.16) also remains bounded. As long as we consider periodic solutions, this means that

$$\int_0^{2\pi} U_\theta H_1 d\theta = 0. \tag{4.17}$$

This is the already familiar orthogonality condition.

After substituting the expression for H_1 from (4.13) into (4.17) we have

$$\frac{\partial}{\partial T}\left(\omega\int_0^\pi U_\theta^2 d\theta\right) + \frac{\partial}{\partial X}\left(k\int_0^\pi U_\theta^2 d\theta\right) = 0. \tag{4.18}$$

As long as the zero-order solution U is known, the PDE (4.18) describes slow variation of parameters of the main solution.

A more specific form of (4.18) can be obtained using the zero approximation (4.12). This equation has the first integral in the form

$$\frac{1}{2}(\omega^2 - k^2)U_\theta^2 + P(U) = E(T, X), \tag{4.19}$$

where E is an integration constant (possibly slowly varying).

The second integration yields an implicit solution:

$$\theta = (\omega^2 - k^2)^{1/2}\int\left[2\left(E - P(U)\right)\right]^{-1/2} dU. \tag{4.20}$$

For the solutions having a period of 2π from (4.20) we obtain the nonlinear dispersion equation, connecting ω, k, and E:

$$(\omega^2 - k^2)^{1/2} \oint \left[2\left(E - P(U)\right)\right]^{-1/2} dU = 1. \tag{4.21}$$

Here the circled integral is taken over one period. In the linear case, when $P(u) = u^2/2$, this produces the dispersion relation

$$\omega^2 = k^2 + 1, \tag{4.22}$$

which is independent of E, i.e., of the wave amplitude.

After using (4.21), Eq. (4.18) acquires the form

$$\begin{aligned} &\frac{\partial}{\partial T}\left(\omega\left(\omega^2 - k^2\right)^{-1/2} \oint \left[2\left(E - P(U)\right)\right]^{1/2} dU\right) \\ &+\frac{\partial}{\partial X}\left(k\left(\omega^2 - k^2\right)^{-1/2} \oint \left[2\left(E - P(U)\right)\right]^{1/2} dU\right) = 0. \end{aligned} \tag{4.23}$$

Equations (4.21) and (4.23) together with the first equation (4.1),

$$\frac{\partial k}{\partial T} + \frac{\partial \omega}{\partial X} = 0, \tag{4.24}$$

form a closed set of equations for slow variations of E, ω, and k.

Finally, it can be readily shown that the same results can be obtained from the averaged variational principle described in Chapter 3. Indeed, the basic equation (4.8) follows from the variation of the action with the Lagrangian

$$L = \frac{1}{2}u_t^2 - \frac{1}{2}u_x^2 - P(u). \tag{4.25}$$

After averaging, we have

$$\mathcal{L}(\omega, k, E) = \left(\omega^2 - k^2\right)^{1/2} \oint \left[2\left(E - V(U)\right)\right]^{1/2} dU - E. \tag{4.26}$$

Variation of this expression with respect to E gives the dispersion equation (4.21), and variation with respect to θ yields the equation

$$\frac{\partial}{\partial T}\mathcal{L}_\omega - \frac{\partial}{\partial X}\mathcal{L}_k = 0, \tag{4.27}$$

which is equivalent to (4.18).

4.2.2. *Attenuation of a periodic wave*

Let us now apply the general results to the specific form of Eq. (4.8):

$$u_{tt} - u_{xx} + u - u^3 = 0, \tag{4.28}$$

The stationary form of this equation, for $u = U(\theta = kx - \omega t)$, is

$$k^2(V^2 - 1)U_{\theta\theta} + U - U^3 = 0, \tag{4.29}$$

where $V = \omega/k$ is the wave velocity. Its solutions are expressed in terms of Jacobi elliptic functions. In particular, for $V > 1$ ("fast" wave) the periodic stationary solution is expressed by the Jacobi elliptic sinus:

$$U = A\mathrm{sn}\left[\frac{2K(s)}{\pi}k(x - Vt), s\right]$$
$$s^2 = \frac{A^2}{2 - A^2}, \quad 0 \le s^2 \le 1. \tag{4.30}$$

Here the parameter s (the modulus) is defined in the interval $0 \le s \le 1$. The functions $K(s)$ and, further, $E(s)$ are the full elliptic integrals:

$$K(s) = \int_0^{\pi/2} \frac{dy}{\sqrt{(1 - s^2\sin^2 y)}}, \quad E(s) = \int_0^{\pi/2} \sqrt{(1 - s^2\sin^2 y)}dy. \tag{4.31}$$

These integrals have an explicit form in the case of small nonlinearity, when $s^2 << 1$. In this case

$$K(s) \approx \frac{\pi}{2}\left(1+\frac{s^2}{4}\right), \quad E(s) \approx \frac{\pi}{2}\left(1-\frac{s^2}{4}\right). \tag{4.32}$$

Another extreme case is that of s being close to 1. Denoting $s' = \sqrt{1-s^2}$, we have

$$K(s) \approx \ln\frac{4}{s'} + \frac{s'^2}{4}\ln\frac{1}{s'}, \quad E(s) \approx 1 + \frac{s'^2}{2}\ln\frac{4}{s'}. \tag{4.33}$$

The solution (4.30) contains two independent parameters: wave velocity V, and its amplitude A (or the nonlinearity parameter s). Note that the period-average of u is zero. The wave amplitude varies within the limits $0 \le A \le A_{max} = 1$, whereas the nonlinear dispersion equation has the form

$$\omega^2 = V^2k^2 = k^2 + \frac{1}{1+s^2}\left(\frac{\pi}{2K(s)}\right)^2. \tag{4.34}$$

Profiles of the wave (4.30) for several values of the nonlinearity parameter s are shown in Fig. 4.2.

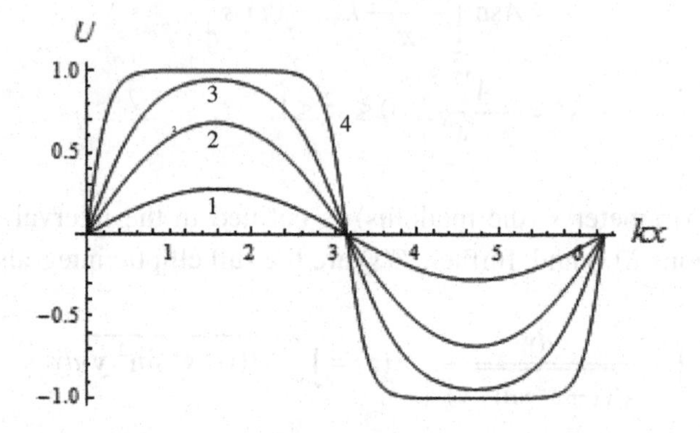

Figure 4.2. Wave forms for $s = 0.04$ (1, close to a sinusoid); 0.3 (2); 0.8 (3); and 0.99 (4) (close to the limiting form).

To describe slow variations of the wave parameters, we substitute (4.30) into (4.18). Note first that

$$Q(s)=\int_0^{\pi} U_\theta^2 d\theta=\frac{4A^2K(s)}{3\pi s^2}\left[\left(1+s^2\right)E(s)-(1-s^2)K(s)\right]. \qquad (4.35)$$

Now Eq. (4.18) has the form

$$\frac{\partial}{\partial T}\left(\omega Q(s)\right)+\frac{\partial}{\partial X}\left(kQ(s)\right)=0. \qquad (4.36)$$

Equations (4.24), (4.34), and (4.36) form a closed system for slowly varying functions. Substituting from (4.30)

$$A^2=\frac{2s^2}{(1+s^2)}, \qquad (4.37)$$

we obtain a system for s and k:

$$\frac{\partial}{\partial T}\left(\frac{\omega Q(s)}{1+s^2}\right)+\frac{\partial}{\partial X}\left(\frac{kQ(s)}{1+s^2}\right)=0,$$

$$\frac{\partial k}{\partial T}+\frac{2k}{\sqrt{4k^2+\pi^2/\left[\left(1+s^2\right)K^2(s)\right]}}\frac{\partial k}{\partial X}$$

$$-\frac{3\pi^3 Q(s)}{16\left(1-s^2\right)\left(1+s^2\right)^2 K(s)\sqrt{4k^2+\pi^2/\left[\left(1+s^2\right)K^2(s)\right]}}\frac{\partial s^2}{\partial X}=0.$$

(4.38)

In the linear limit, $s\to 0$, these equations yield

$$\begin{aligned} &\left(\omega A^2\right)_T + \left(c_g \omega A^2\right)_X = 0, \\ &k_T + c_g k_X = 0. \end{aligned} \tag{4.39}$$

Here, $c_g = d\omega / dk = k / \sqrt{1+k^2}$ is the linear group velocity. The first equation (4.39) reflects energy conservation, and the second follows directly from (4.24) for a given dispersion equation $\omega = \omega(k)$.

Now we add a small dissipation to (4.28); namely,

$$u_{tt} - u_{xx} + u - u^3 = -\mu \nu u_t, \quad \nu > 0. \tag{4.40}$$

Then in the first approximation the term $-\nu\omega U_\theta$ should be added to the function H_1. For simplicity we suppose that the wave train is uniform in space so that its parameters do not depend on X but vary with T. As a result we have

$$\frac{\partial}{\partial T}\left(\frac{\omega Q(s)}{1+s^2}\right) = -\nu \frac{\omega Q(s)}{1+s^2}, \tag{4.41}$$

where ω and s are functions of T. From here it immediately follows that

$$\frac{\omega Q(s)}{1+s^2} = \text{const} \cdot e^{-\nu T}. \tag{4.42}$$

In particular, for a linear wave ($s \ll 1$), from (4.35) and (4.37) it follows that $Q = \pi s^2 = \pi A^2 / 2$, and the left-hand side of Eq. (4.42) reduces to

$$\frac{\omega Q(s)}{1+s^2} \approx \left(\frac{\pi}{2}\sqrt{1+k^2}\right) s^2, \tag{4.43}$$

which results in

$$s \sim A \sim e^{-\nu T/2}. \tag{4.44}$$

For the general case of Eq. (4.42), Fig. 4.3 shows the time dependence of slowly varying wave parameters.

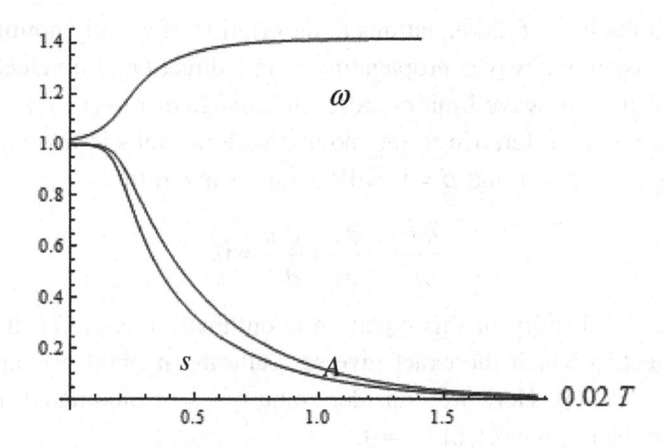

Figure 4.3. Variation of nonlinearity parameter s, wave amplitude A, and frequency ω, in a damping wave with $k = 1$, at $\nu = 0.1$. Initial conditions are $s(0) = 0.99$ and $A(0) = 1$.

Note that wave damping results in changing of the wave period (basic frequency) due to its dependence on the wave amplitude. The corresponding variation of the wave form can be seen from Fig. 4.2 in which lines from 4 down to 1 are now pertinent to $T = 0$, 9.5, 16.7, and 34.2, respectively.

4.3. Korteweg–de Vries Equation

4.3.1. *Stationary waves*

Another important example is the Korteweg–de Vries (KdV) equation, one of the most popular in nonlinear wave theory. Its general form is

$$\frac{\partial u}{\partial t} + c_0 \frac{\partial u}{\partial x} + \alpha u \frac{\partial u}{\partial x} + \beta \frac{\partial^3 u}{\partial x^3} = 0, \tag{4.45}$$

where c_0, α, and β are constant parameters. This equation was first derived by Korteweg and de Vries in 1895 for waves on the water surface, under the assumption that the terms with α (nonlinearity) and β (dispersion) are small in comparison with each of the first two terms. Since then, KdV was derived in numerous areas of physics as a result of reducing the basic field equations to description of weakly nonlinear and weakly dispersive waves propagating in one direction at a velocity close to its linear, long-wave limit c_0. After the change of x to $x - c_0 t$ (Galilean transform to the reference frame moving with the velocity c_0) and letting, for simplicity, $\alpha = 1$ and $\beta = 1$, KdV acquires the form

$$\frac{\partial u}{\partial t} + u\frac{\partial u}{\partial x} + \frac{\partial^3 u}{\partial x^3} = 0. \qquad (4.46)$$

The rich history of this equation is outlined in, e.g., [7]. It was the first object to which the exact inverse scattering method was applied in 1967 (e.g., [11]). Here we consider some types of modulated nonlinear waves in the framework of (4.46).

As always, we begin with the stationary traveling waves. After substitution of $u = U(\theta = kx - \omega t)$ and integration, we have an ODE:

$$k^2 \frac{d^2U}{d\theta^2} + \frac{1}{2}U^2 - VU = B, \qquad (4.47)$$

with $B = \text{const}$ and $\omega = kV$.

To obtain an idea of the family of solutions of this equation, Fig. 4.4 shows its phase plane showing phase trajectories, i.e., the dependencies between U and $dU/d\theta$, for different initial conditions.

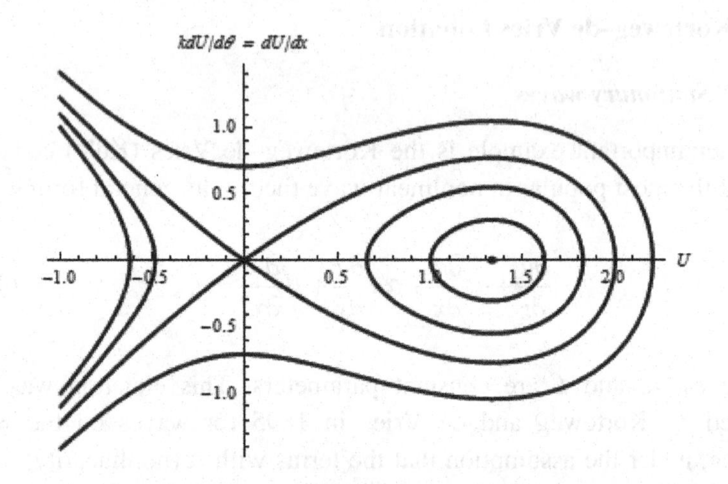

Figure 4.4. Phase trajectories for the family of stationary solutions of Eq. (4.47) for the case $B = 0$, so that $U = 0$ is one of the equilibrium points.

The corresponding solutions for U are again expressed via Jacobi elliptic functions. In what follows we represent the solution of (4.47) in the form taken in papers [3, 4] which will be discussed below:

$$U = \gamma + \left(\frac{2A}{s^2}\right) dn^2\left(\sqrt{\frac{A}{6k^2s^2}}\theta, s\right), \tag{4.48}$$

where γ and s are constants. The function $dn(x, s)$ is one of the Jacobi elliptic functions (delta-amplitude). As in (4.30), the parameter s (the modulus) is defined in the interval $0 \le s \le 1$. The functions $K(s)$ and $E(s)$ are the full elliptic integrals defined by (4.31).

In another, equivalent presentation (elliptic functions are related to each other), the solutions (4.48) are expressed in terms of elliptic cosine, *cn*, and they are often called *cnoidal waves* (the wave in the *dn* representation is sometimes called “dnoidal wave”).

The wave amplitude $A = (U_{max} - U_{min})/2$, wave number k, and velocity $V = \omega/k$ are related as

$$V = \frac{\omega}{k} = \gamma + 2A\left(\frac{2-s^2}{3s^2}\right), \quad k = \frac{\pi}{K(s)}\sqrt{\frac{2A}{6s^2}}, \tag{4.49}$$

and the period-average of the solution is

$$\langle U \rangle = \gamma + \frac{2AE(s)}{s^2K(s)}. \tag{4.50}$$

Figure 4.5 shows the function dn^2 at different s. At small s the wave form is close to a sinusoid, whereas at s close to unity, it becomes a series of impulses, each close to a solitary wave, a soliton. We shall specially discuss the dynamics of solitons later in this book.

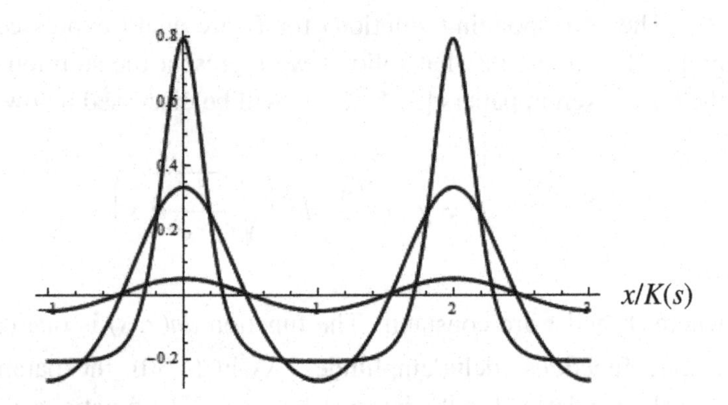

Figure 4.5. The function $dn^2(x,s)-[E(s)/K(s)]$ (its period average is zero) at s = 0.1, 0.6, and 0.999.

4.3.2. *Slowly varying cnoidal waves*

The KdV equation can be written in Lagrangian form after which the averaged Euler–Lagrange equations can be used to describe its solutions with slowly varying parameters. Several examples were considered in the literature. In [8], evolution of a cnoidal water wave propagating over a sloping bottom was considered; upon reaching a shallower area, an almost sinusoidal wave with small modulus s can be transformed into a train of pulses close to solitons (s close to 1). In [9], damping of a cnoidal wave due to dissipative factors is considered; in this case a wave with s close to 1 can eventually become close to a sinusoid, similarly to what was considered above for the Klein–Gordon equation. Damping of a soliton will be considered in Chapter 6 below.

Note that, due to the nature of the KdV equation as a weakly nonlinear and weakly dispersive limit of basic physical equations (this is true for many other "evolution" equations in the form of $\partial u/\partial t = F(u)$, where F is a differential operator containing x-derivatives), the basic approximation of the theory can be used in a simplified way. In general, the perturbed KdV equation with coefficients slowly varying in space has the form

$$\frac{\partial u}{\partial x}+c_0^{-1}(X)\frac{\partial u}{\partial t}+\alpha_1(X)u\frac{\partial u}{\partial t}+\beta_1(X)\frac{\partial^3 u}{\partial t^3}=\mu R(u,X). \quad (4.51)$$

Here the operator R is still much smaller than the terms with α_1 and β_1 so that the latter can be of the order of $\sqrt{\mu}$. This equation is written in the so-called "time-like" form which is more convenient for consideration of spatial variations of the wave parameters, such as propagation in an inhomogeneous medium or cylindrical divergence/convergence of the wave. If the wave is modulated in space but remains strictly periodic in time, Eq. (3.45) of Chapter 3 for the average Lagrangian reduces to

$$\frac{d}{dX}\left(\frac{\partial \mathcal{L}}{\partial k}\right)=\frac{\partial\langle Q\rangle}{\partial\omega}, \quad (4.52)$$

or, according to (3.41),

$$\frac{\partial\langle S\rangle}{\partial X}=-\frac{1}{\omega}\frac{\partial\langle Q\rangle}{\partial\omega}, \quad (4.53)$$

where, again, $R=-\partial Q/\partial U_t$ and S is energy flux averaged over the wave period. For (4.51) in the main approximation, the functions characterizing the wave energy and momentum can be represented as the same quadratic functions of the variable u as they are defined in the linear, non-dispersive approximation. As a result, Eq. (4.53) becomes equivalent to that obtained by multiplying (4.51) by u and integrating over the time period at fixed coefficients:

$$\frac{d}{dX}\langle u^2\rangle=2\langle uR\rangle. \quad (4.54)$$

We shall consider wave dissipation in more detail in Chapter 6 in application to a soliton. In the rest of this chapter we shall discuss simple envelope waves in the KdV and their application to evolution of a stepwise initial wave.

4.3.3. *Conservation equations*

Obtaining the averaged equations for the KdV is, in general, a cumbersome task. This was done by Whitham [10]; here we reproduce the main points of his derivation. Whitham considered the KdV equation (4.46) (with a factor 6 at the nonlinear term uu_x, so we have adjusted his results to the form (4.46)) and derived three conservation equations:

$$\begin{aligned} &u_t + (u^2/2 + u_{xx})_x = 0, \\ &\left(u^2/2\right)_t + \left(u^3/3 + uu_{xx} - u_x^2/2\right)_x = 0, \\ &\left(u^3 - u_x^2/2\right)_t + \left(3u^4/4 + 3u^2 u_{xx} + u_{xx}^2/2 + u_x u_t\right)_x = 0. \end{aligned} \tag{4.55}$$

The first of these equations is simply equivalent to (4.46), the second follows from multiplying (4.46) by u, and the third, from multiplying it by u^2 and using (4.46) again. These equations can be easily verified. They can be interpreted as those expressing conservation of mass, energy, and momentum with respect to Eq. (4.46) (but not necessarily for the basic physical equations for which KdV represents an approximation).

Now the averaged equations can be written. For this, an auxiliary, slowly varying function defined on the stationary solutions $U(X, T, \theta)$ is introduced:

$$W(A,B,V) = -k\oint U_\theta dU = -k\sqrt{2}\oint\sqrt{\left(-A + BU + VU^2/2 - U^3/6\right)}dU. \tag{4.56}$$

Here A and B are new constants defined from the first integral of (4.47):

$$\frac{k^2}{2}U_\theta^2 = -A + BU + \frac{1}{2}kVU^2 - k\frac{U^3}{6}. \tag{4.57}$$

Note that the derivative W_A defines the wave spatial period, so that $k = 1/W_A$ is the basic wave number.

Differentiating W with respect to A, B, and V and substituting into (4.55), we obtain

$$\begin{aligned}
&\frac{\partial}{\partial T}(kW_B)+\frac{\partial}{\partial X}(kVW_B - B)=0,\\
&\frac{\partial}{\partial T}(kW_V)+\frac{\partial}{\partial X}(kVW_V - A)=0,\\
&\frac{\partial}{\partial T}\left[k\left(AW_A+BW_B+VW_V-W\right)\right]\\
&+\frac{\partial}{\partial X}\left[kV\left(AW_A+BW_B+VW_V-W\right)-B^2/2-AV\right]=0.
\end{aligned} \tag{4.58}$$

Together with the continuity equation (4.24),

$$\frac{\partial k}{\partial T}+\frac{\partial (Vk)}{\partial X}=0, \tag{4.59}$$

the system (4.58) simplifies to

$$\begin{aligned}
&\left(\frac{\partial}{\partial T}+V\frac{\partial}{\partial X}\right)(W_A)-W_A\frac{\partial V}{\partial X}=0,\\
&\left(\frac{\partial}{\partial T}+V\frac{\partial}{\partial X}\right)(W_B)-W_A\frac{\partial B}{\partial X}=0,\\
&\left(\frac{\partial}{\partial T}+V\frac{\partial}{\partial X}\right)(W_V)-W_A\frac{\partial A}{\partial X}=0.
\end{aligned} \tag{4.60}$$

Calculation of characteristic velocities in this hyperbolic system still remains cumbersome. Whitham found them in terms of three roots, $r_{1,2,3}$, of the polynomial under the square root in Eq. (4.56). (In what follows we use the notations from [3, 4].) The parameters of the solution (4.48) are related to these variables (ranged as $r_3 \geq r_2 \geq r_1$) as

$$a=r_2-r_1,\quad s^2=\frac{r_2-r_1}{r_3-r_1},\quad \gamma=r_1+r_2-r_3. \tag{4.61}$$

Then we have

$$V = \frac{1}{3}(r_1 + r_2 + r_3),$$
$$U_{\max} = r_3 + r_2 - r_1,\ u_{\min} = r_3 + r_1 - r_2. \tag{4.62}$$

In these variables, Eqs (4.60) can be written in the form

$$\frac{\partial r_\alpha}{\partial T} + V_\alpha \frac{\partial r_\alpha}{\partial X} = 0,\ \alpha\text{=}1,2,3. \tag{4.63}$$

The velocities V_α depend on the variables r_α and s:

$$V_1 = \frac{1}{3}\left[(r_1 + r_2 + r_3) - \frac{2(r_2 - r_1)K(s)}{K(s) - E(s)}\right],$$
$$V_2 = \frac{1}{3}\left[(r_1 + r_2 + r_3) - \frac{2(r_2 - r_1)(1 - s^2)K(s)}{E(s) - (1 - s^2)K(s)}\right], \tag{4.64}$$
$$V_3 = \frac{1}{3}\left[(r_1 + r_2 + r_3) + \frac{2(r_3 - r_1)(1 - s^2)K(s)}{E(s)}\right].$$

4.3.4. *Evolution of step function*

Based on Whitham's approach, Gurevich and Pitaevskii [3, 4] considered wave evolution from the initial condition in the form of a stepwise perturbation:

$$u(t = 0) = \begin{Bmatrix} 1, & x < 0, \\ 0, & x > 0 \end{Bmatrix}. \tag{4.65}$$

The solution is sought for large t when, due to the dispersion, the wave acquires the form of an oscillating train. For this stage, we look for the self-similar, simple envelope waves already discussed in Section 4.2. We consider all r_α as functions of $\xi = X / T$, to obtain from (4.64)

$$(V_\alpha - \xi)\frac{dr_\alpha}{d\xi} = 0. \tag{4.66}$$

Thus, if one of r_α in (4.63) is non-constant, the corresponding velocity should satisfy Eq. (4.7), $V_\alpha = X/T$, whereas in the remaining two equations (4.63), r_α should be constants.

To specify the solution, one should add boundary conditions. It can be supposed (and confirmed by calculations), that asymptotically, at the front of the wave train (denoted by subscript +) a soliton will eventually be formed, so that

$$s = 1,\ r_3 = r_2 = r_+,\ \langle U \rangle = 0,\ \xi = \xi_+, \tag{4.67}$$

and at the rear end of the train (marked by –) the wave is linear on the background of the constant:

$$s = 0,\ a = 0,\ r_2 = r_1 = r_-,\ \langle U \rangle = 1,\ \xi = \xi_-. \tag{4.68}$$

To satisfy the boundary conditions (4.67), we have to let $\gamma = r_1 = 0$ and $a = r_2$, so that if $a \neq const$, one should let $V_2 = \xi$ and r_3 = const. Then, using the last equality in (4.68) and the expression (4.50) for $\langle U \rangle$, we have $r_3 = 1$.

At the other end, from (4.68) we have $a/s^2 = r_3$ and $\gamma = -1$. Collecting all these results, we obtain

$$a = Bs^2,\ V = B(1+s^2)/3,\ \gamma = -(1-s^2), \tag{4.69}$$

And finally, from (4.64), we find that

$$V_2 = \frac{1}{3}\left\{1 + s^2 - \frac{2s^2\left(1-s^2\right)K(s)}{E(s) - \left(1-s^2\right)K(s)}\right\} = \frac{X}{T}. \tag{4.70}$$

Now we have a closed system of equations for the "envelope waves."

The authors of [3, 4] then analyzed the solution by considering its frontal and trailing edges, using the expansions (4.32) and (4.33) for the elliptic integrals. In such a wave, the trailing edge is a quasi-harmonic wave in which

$$a = s^2 = (2/3)\left[(X/T)+1\right]. \tag{4.71}$$

From here, the amplitude of oscillations tends to zero at $X = -2T/3$; on the left of this (moving) point there are no perturbations and, in this approximation, u = const. The frontal zone of the wave is close to a sequence of solitons, so that s^2 is close to 1, and

$$(1-s^2)\ln\frac{16}{1-s^2} = 2-3(X/T). \tag{4.72}$$

This wave exists at $X \le 2T$, and the perturbed interval is linearly expanding with time as $(5/3)T$. The resulting wave profiles are shown in Fig. 4.6.

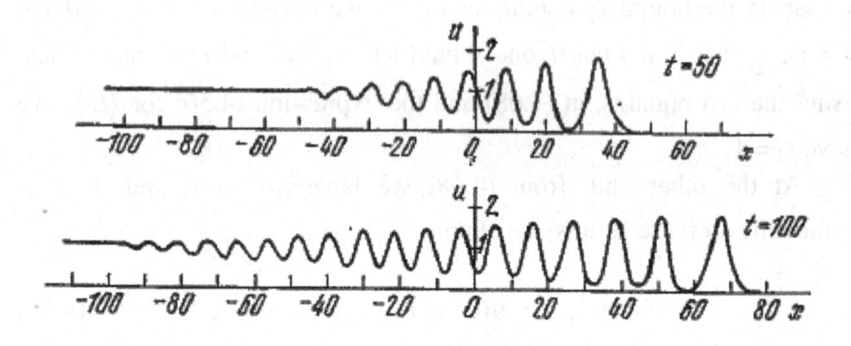

Figure 4.6. The form of an initially stepwise wave at two time moments. From [5, 6].

In [4] this solution is considered as a model of the so-called collisionless shock waves in plasma. Note also that in the 1960s, electromagnetic shock waves in the form of oscillating fronts were observed in nonlinear magnetic materials (ferrites) [2].

4.4. Concluding Remarks

In this chapter we demonstrated how the perturbation method for non-harmonic waves can work for two widely used nonlinear equations, Klein–Gordon and Korteweg–de Vries. One of the interesting processes is a slow transition between a sequence of kinks or solitons ($1 - s^2 << 1$) and almost linear (quasi-harmonic) wave profile ($s << 1$). This approach has numerous applications in plasma physics, water wave theory, and electrodynamics. The same approach worked for other nonlinear equations, such as MKdV and Gardner equations, considered later in this book in relation to solitons and kinks.

References

1. Courant, R. and Friedrichs, K. O. (1948). *Supersonic Motion and Shock Waves.* Springer, New York and Berlin.
2. Gaponov, A. V., Ostrovsky, L. A., and Freidman, G. I. (1967). Shock waves. *Radiophysics and Quantum Electronics, v.* 10, pp. 772–793.
3. Gurevich, A. V. and Pitaevskii, L. P. (1973). Decay of initial discontinuity Korteweg–de Vries equation. *JETP Letters,* v. 17, pp. 193–195.
4. Gurevich, A. V. and Pitaevskii, L. P. (1974). Nonstationary structure of a collisionless shock wave. *Soviet Physics JETP*, 38, pp. 291–297.
5. Landau, L. D. and Lifshits, E. M. (1987). *Fluid Mechanics.* Pergamon Press, Oxford.
6. Luke, J. C. (1966). Perturbation method for nonlinear dispersive wave problems. Proceedings of the Royal Society London, v. 292, pp. 403–412.
7. Miles, J. W. (1981). The Korteweg–de Vries equation: a historical essay. *Journal of Fluid Mechanics*, v.106, pp. 131–147.
8. Ostrovsky, L. A. and Pelinovsky, E. N. (1970). Wave transformation on the surface of a fluid of variable depth. *Izvestija Atmospheric and Oceanic Physics*, v. 6, pp. 934–939.
9. Pelinovsky, E. N. (1971). On the absorption of nonlinear waves by dispersive media *Journal of Applied Mechanics and Technical Physics,* v. 12, pp. 227–230.
10. Whitham, G. B. (1965). Non-linear dispersive waves. *Proceedings of the Royal Society London,* v. 283, pp. 238–261.
11. Whitham, G. B (1974) *Linear and Nonlinear Waves*. J. Wiley & Sons, New York.

Chapter 5

Perturbation Methods for Solitary Waves and Fronts

> Some scientists work so hard there is no time left for serious thinking.
>
> Francis Crick

In the previous chapters we expanded our consideration from quasi-harmonic waves to nonlinear quasi-periodic non-sinusoidal waves. Now we make another step and consider, instead of periodic waves, the class of localized waves which tend to constants (not necessarily zeros) at infinity. This class includes conservative solitary waves such as a soliton and a kink. However, the scheme discussed below can work for dissipative and active systems, supporting shock waves and autowaves.

As above, the basic, non-perturbed solution is a stationary traveling wave described by an ODE. In this chapter such a solution is a limiting trajectory, a separatrix, in the phase space (in particular, phase plane) of these stationary equations. We already touched upon this topic in Chapter 4 when considering a quasi-periodic "cnoidal" wave which tends to a series of solitons when the nonlinearity parameter s approaches unity. Now we consider solitons and other localized waves as separate, slowly varying entities.

The contents of this chapter needs an additional explanation. We begin from the direct, "quasi-stationary" perturbation scheme similar to that considered in Chapter 3 but significantly modified to describe localized waves and their "radiation". Considering that in the last few decades solitons have been one of the most intensively studied objects of nonlinear theory, we then briefly outline another, more sophisticated perturbation scheme, applicable to solitons in systems close to exactly integrable ones. Nonetheless, in this book we mainly explore the relatively simple, direct approach which is applicable to both integrable and non-integrable equations of a broader class, including dissipative and active systems.

5.1. Quasi-Stationary Theory

Consider a direct perturbation scheme [5] which is somewhat similar to that developed in Chapter 3 for quasi-periodic, non-harmonic waves. The difference from the previous schemes is that now one must avoid secular growth of perturbations at large distances from the localized wave, both in front of and behind it, rather than accumulation of perturbations from period to period. Another important point is that, in higher approximations, a small perturbation can spread outside the localized wave as "radiation". To clarify this, below we describe the perturbation scheme with the corresponding modifications.

Consider again a system of N nonlinear equations:

$$\mathbf{M}(\mathbf{u},T,\mathbf{X}) = A(\mathbf{u},T,\mathbf{X})\frac{\partial \mathbf{u}}{\partial t} + \mathbf{B}(\mathbf{u},T,\mathbf{X})\frac{\partial \mathbf{u}}{\partial \mathbf{r}} + C(\mathbf{u},T,\mathbf{X}) = \mu \mathbf{f}(\mathbf{u},T,\mathbf{X}). \quad (5.1)$$

Here, as in Chapter 3, $\boldsymbol{u} = \{u_1, u_2, \ldots, u_N\}$ is a vector of dependent variables, $T = \mu t$ and $\mathbf{X} = \mu \mathbf{r}$ are "slow" variables, and $\mu << 1$ is a small parameter. It is assumed that at $\mu = 0$ the system (5.1) has a particular solution in the form of a stationary plane wave $\mathbf{u}^{(0)} = \mathbf{U}(\zeta, \mathbf{A})$, where $\zeta = t - \mathbf{qr}$, and $\mathbf{A} = \{A_1, A_2, \ldots, A_m\}$ is a set of integration parameters (they can be the solitary wave coordinates, velocities, or amplitudes). Wave velocity is $\mathbf{V} = \mathbf{q}/q^2$, so that the vector $\mathbf{q}$ can be called the "slowness vector". Far from the wave center these solutions tend to constants, $\mathbf{U}_-$ at $\zeta \to -\infty$ and $\mathbf{U}_+$ at $\zeta \to +\infty$ (i.e., far before and far after the solitary wave front). These localized stationary solutions satisfy the ordinary equations:

$$\left[A^{(0)}(\mathbf{U})-\mathbf{qB}^{(0)}(\mathbf{U})\right]\frac{\partial \mathbf{U}}{\partial \zeta}+C^{(0)}(\mathbf{U})=0, \tag{5.2}$$

where the superscript (0) denotes zero approximation. As mentioned, each solution with the above constant asymptotics forms a separatrix (or, in general, a separatrix hypersurface) in the phase space of the system (5.2) which connects $\mathbf{U}_-$ and $\mathbf{U}_+$. In a "classic" soliton, $\mathbf{U}_- = \mathbf{U}_+$, but in general these constants are not necessarily zeros and can differ from each other. For example, this solution can be a *kink*, a transition between two different constant asymptotics.

At $\mu \neq 0$ the solution in the vicinity of a given solitary wave $\mathbf{U}(\zeta,\mathbf{A})$ is again represented by a series:

$$\mathbf{u}(\mathbf{r},t)=\mathbf{U}(\zeta,T,\mathbf{X})+\sum_{n=1}^{J}\mu^n\mathbf{u}^{(n)}(\zeta,T,\mathbf{X}), \tag{5.3}$$

where J is an integer which defines the order of the approximation. Now $\zeta = t-\int \mathbf{q}(T,\mathbf{X})d\mathbf{r}$, the parameters $\mathbf{A}$ and $\mathbf{q}$ are assumed to be functions of slow variables T and $\mathbf{X}$. The wave velocity is $\mathbf{V}=d\mathbf{X}/dT$, so that the integral in ζ is taken along the wave trajectory, and the dependencies on T and $\mathbf{X}$ are related. Then, as before, we substitute the series (5.3) into Eqs (5.1) and expand the resultant equations in powers of μ. As a result, the set of linear ODE follows in each n-th order:

$$G\mathbf{u}^{(n)}=\mathbf{H}^{(n)}, \text{ with}$$

$$G=\left(A^{(0)}-\mathbf{qB}^{(0)}\right)\frac{d}{d\zeta}+\left(\frac{\partial A^{(0)}}{\partial \mathbf{U}}-\mathbf{q}\frac{\partial \mathbf{B}^{(0)}}{\partial \mathbf{U}}\right)\frac{\partial \mathbf{U}}{\partial \zeta}+\frac{\partial C^{(0)}}{\partial \mathbf{U}}, \tag{5.4}$$

and the vectors $\mathbf{H}^{(n)}$, as above, contain functions from previous approximations. In particular,

$$\mathbf{H}^{(1)}=\mathbf{f}^{(0)}-A^{(0)}\frac{\partial \mathbf{U}}{\partial T}-\mathbf{B}^{(0)}\frac{\partial \mathbf{U}}{\partial \mathbf{X}}. \tag{5.5}$$

The general solution of (5.4) in each approximation has the same form as in Chapter 3:

$$\mathbf{u}^{(n)} = Y\left(\mathbf{C}^{(n)} + \int_0^{\zeta} Y^{*}\mathbf{H}^{(n)} d\zeta'\right), \tag{5.6}$$

where $\mathbf{C}^{(n)}(T, \mathbf{X})$ is the integration constant (independent of ζ); Y is the matrix of fundamental solutions of the homogeneous set of equations $GY = 0$; and Y^{*} is the adjoint matrix. Again, $m + 1$ columns of matrix Y can be found from the basic solution $\mathbf{U}(\zeta, \mathbf{A})$ by variation of Eq. (5.2) with respect to ζ and A_i and comparison with (5.4):

$$\mathbf{Y}_1 = \mathbf{U}_{\xi}; \quad \mathbf{Y}_{i+1} = \mathbf{U}_{A_i}; \quad i = 1, 2, ..., m. \tag{5.7}$$

Thus far we simply reproduced the scheme described in Chapter 3 for quasi-periodic waves. Now we shall address the solitary waves' specifics by requiring that small perturbations remain bounded at $\zeta \to \pm\infty$. For that, asymptotic behavior of perturbation (5.6) should be analyzed. All considered functions are supposed sufficiently smooth to secure the regular asymptotic behavior.

Some vectors $\mathbf{Y}_j$ tend to zero at $\zeta \to \pm\infty$. These are the vector $\mathbf{Y}_1$ and those vectors $\mathbf{Y}_{i+1} = \mathbf{U}_{A_i}$ ($i = 1, 2, \ldots, l$) for which $\partial\mathbf{U}_{\pm}/\partial A_i = 0$ (we will refer to the corresponding A_i as *internal parameters* of the wave). The other vectors $\mathbf{Y}_{i+1} = \mathbf{U}_{A_i}$ ($i = l + 1, \ldots, m$) approach non-zero constants when $\zeta \to \pm\infty$ (we refer to the corresponding quantities A_i as *pedestals*). The remaining $N - (m + 1)$ vectors asymptotically tend to exponents:

$$\mathbf{Y}_i(\zeta \to \pm\infty) \sim e^{\lambda_i \zeta}, \quad i = m+2, m+3, ..., N. \tag{5.8}$$

Indeed, if the basic solution tends to a constant fast enough, asymptotically the linear system (5.4) has constant coefficients, and its asymptotic solution should consist of exponents. The parameters λ_i are, in general, complex, whereas for the solutions (5.7), $\lambda_i = 0$.

The adjoint matrix Y^* possesses similar properties. Some of the solutions corresponding to (5.7), say $m-l$ of them, tend to constants, and other l exponentially grow as $\left(\mathbf{U}_\zeta\right)^{-1}$ when $\zeta \to \pm\infty$.

Consider now the conditions necessary for the solutions (5.6) to remain bounded. If, among the fundamental vectors, there are solutions $\mathbf{Y}_\alpha$ which grow exponentially when $\zeta \to \pm\infty$ (the corresponding $\mathbf{Y}_\alpha^*$ exponentially vanish), then the boundedness of the corresponding terms in (5.6) can be achieved either by choosing the constants $\mathbf{C}_\alpha$:

$$\mathbf{C}_\alpha^{(n)} = -\int_0^\infty \mathbf{Y}_\alpha^* \mathbf{H}^{(n)} d\zeta \,, \tag{5.9}$$

if the integrand $\mathbf{Y}_\alpha^* \mathbf{H}^{(n)}$ is odd, or by the condition

$$\int_0^\infty \mathbf{Y}_\alpha^* \mathbf{H}^{(n)} d\zeta = 0 \,, \tag{5.10}$$

if the integrand is even.

If there exists a fundamental solution $\mathbf{Y}_\beta$ which grows exponentially only in one asymptotic limit, at $\zeta \to +\infty$ or $-\infty$, and vanishes in another limit, then the boundedness of the corresponding terms in (5.6) can be achieved by imposing the condition (5.9).

Finally, perturbation divergence may occur when some solutions (5.7) are finite but non-zero at $\zeta \to \pm\infty$, which leads to a secular growth of $\mathbf{u}^{(n)}$. To suppress such a divergence, one should require that

$$\lim_{\zeta \to \pm\infty} \mathbf{Y}_i^* \mathbf{H}^{(n)} = 0, \quad i = l+1, l+2, \ldots, m+1 \,. \tag{5.11}$$

If the conditions (5.9)–(5.11) are met, the perturbations $\mathbf{u}^{(n)}$ are bounded at all ζ. Note that the character of perturbations described by individual terms in Eq. (5.6) can be qualitatively different. Perturbations associated with the vectors $\mathbf{Y}_\alpha$ which increase to infinity or decrease to zero at $\zeta \to \pm\infty$, are spatially localized; these perturbations vanish asymptotically when $\zeta \to \pm\infty$ and, hence, describe the disturbances of the wave structure in the vicinity of its center. Perturbations associated with non-vanishing vectors $\mathbf{Y}_i$ $(i=l+1,...m+1)$, remain finite at infinity and correspond to large-scale perturbations (their spatial length is much greater than the characteristic scale of the wave); they can be matched with a slowly varying "external" field and, in particular, can be considered as a small radiation.

In the first approximation the integral conditions (5.10) determine the dependence of the internal parameters $\mathbf{q}$ and A_i ($i = 2, 3, \ldots, l$) on T and $\mathbf{X}$. Algebraic conditions (5.11) for $n = 1$ provide an independent description of "pedestals" A_i ($i = l + 1, l + 2, \ldots, m + 1$) and, hence, of the the wave asymptotics $\mathbf{U}_{\pm}(\mathbf{A})$. In higher approximations the integral conditions can be satisfied by means of constants $\mathbf{C}_i^{(n)}$ ($i = 1, 2, \ldots, l$), which remain indefinite in previous approximations (it is assumed that the number of such constants is equal to the number of integral conditions). Note that if the number of eigenvectors $\mathbf{Y}_i$ growing at $\zeta \to \pm\infty$ is equal to the number of vanishing eigenvectors, a bounded solution can be constructed in any approximation.

Secular divergence of the higher-order solutions, at $n > 1$, can be suppressed in the following way. As mentioned, the conditions (5.11) provide equations for possible non-localized components of perturbations; we denote them $\overline{\mathbf{u}}^{(n)}$. By finding these components separately for $\zeta \to +\infty$ and $\zeta \to -\infty$, one can satisfy the conditions (5.11) for each n. For the construction of a solution continuous at all ζ, Eqs (5.11) should be augmented by the matching conditions for $\overline{\mathbf{u}}_+^{(n)}$ and $\overline{\mathbf{u}}_-^{(n)}$ at $\zeta \to 0$ (more specifically, at an "internal" spatial scale of the order of the characteristic width of the localized wave). These matching conditions can be obtained from the analysis of the solutions $\overline{\mathbf{u}}^{(n)}$ in the range $\lambda^{-1} \ll |x| \ll (\mu\lambda)^{-1}$, where λ is the "internal" scale of the wave and $x = 0$ is the position of the wave center at a given moment. In this range the parameters $\mathbf{C}_i^{(n)}$ are constants by definition, so that the perturbation (5.6) changes by $\int_{-\infty}^{+\infty} \mathbf{Y}_i^* \mathbf{H}^{(n)} dz$. In other words, the radiation field $\overline{\mathbf{u}}^{(n)}$ experiences a "jump" in the vicinity of the wave center:

$$\left[\overline{\mathbf{u}}^{(n)}\right] \equiv \overline{\mathbf{u}}_+^{(n)}(0) - \overline{\mathbf{u}}_-^{(n)}(0) = \mathbf{Y}_i \int_{-\infty}^{+\infty} \mathbf{Y}_i^* \mathbf{H}^{(n)} d\zeta, \quad i = l+1, l+2, ..., m+1. \tag{5.12}$$

This finalizes the formal algorithm of the solution.

The procedure described above on the one hand represents a modification of the perturbation scheme for quasi-periodic waves considered in Chapter 3, and on the other hand, it is in many aspects analogous to the method of matched asymptotic expansions. Indeed, the general solution contains a range of "fast" field variation (the internal range) and a range of "slow" field variation (the external range). The difference is that in our case the structure of the internal solution is already known in the zero approximation: it is nothing but the stationary localized wave $\mathbf{U}(\zeta, \mathbf{A})$, and only its slowly varying parameters are to be defined. According to the matched asymptotic expansions method, the main problem is matching the solutions found separately in the internal and external ranges. In the scheme described above, the matching condition is met identically if Eqs (5.9)–(5.11) are satisfied. In particular, in the main approximation the matching of internal and external solutions is provided simply by the form of stationary solution $\mathbf{U}(\zeta, \mathbf{A})$. In higher approximations, the matching is performed for the corresponding radiation components of perturbations (5.6).

5.2. Lagrangian Description

Continuing an analogy with Chapter 3, we now briefly consider the field equations in the Lagrangian form, (3.24):

$$\frac{\partial}{\partial t}\frac{\partial L}{\partial \mathbf{u}_t} + \frac{\partial}{\partial \mathbf{r}}\frac{\partial L}{\partial \mathbf{u}_\mathbf{r}} - \frac{\partial L}{\partial \mathbf{u}} = \mu \mathbf{R}, \tag{5.13}$$

where L is the Lagrangian density. We assume again that at $\mu = 0$, Eqs (5.13) have a particular family of solutions in the form of a localized wave, $\mathbf{U}(\zeta, \mathbf{A})$, satisfying the system of ODEs:

$$\frac{d}{d\zeta}\left(\frac{\partial L}{\partial \mathbf{U}_t} - \mathbf{q}\frac{\partial L}{\partial \mathbf{U}_\mathbf{r}}\right) - \frac{\partial L}{\partial \mathbf{U}} = 0. \tag{5.14}$$

We again look for the solution in the form of a series (5.3) so that in each order of μ we obtain a linear inhomogeneous ODE system similar to Eqs (3.29) of Chapter 3. As above, some of the solutions of the homogeneous system $G_L Y = 0$ are known as the derivatives of $\mathbf{U}$ with respect to ζ and the other possible free parameters in $\mathbf{U}$. Moreover, similar to that in (3.29), this operator is self-adjoint, which means that at least one fundamental solution of the equation $G_L Y^* = 0$ vanishing at $\zeta \to \pm\infty$, namely $\mathbf{U}_\zeta$, is known. Thus, in each approximation it is necessary to meet at least the following orthogonality condition:

$$\int_{-\infty}^{\infty} \mathbf{U}_\zeta \mathbf{H}^{(n)} d\zeta = 0. \tag{5.15}$$

In some interesting cases (such as the interaction of solitons with close velocities to be considered later), variation of wave velocity is not only slow but also small. In this case (for simplicity, in the one-dimensional geometry) the basic solution $\mathbf{U}$ depends on the variable ζ = $t - x/c_0 - S_i(T)$, where c_0 is the unperturbed linear velocity and S_i is the slowly varying local coordinate of the i-th soliton in the reference frame moving with the velocity c_0. In this case the first-order perturbation is

$$\mathbf{u}^{(1)} = (\partial \mathbf{U} / \partial c_0)(dS_i / dT). \tag{5.16}$$

This function is always finite, so that the divergence can arise only in the second-order approximation.

5.3. Non-Stationary Perturbations

As mentioned, the asymptotic scheme for localized waves described above is similar to that developed in the previous chapters for quasi-periodic waves. In these cases perturbations have the same structure as the basic solution, namely, they depend on only one "fast" variable ζ and possibly on the slow variables T and $\mathbf{X}$. That was a reasonable simplification, since our goal was to prevent secular growth of perturbations due to resonance with the "force" $\mathbf{H}^{(n)}$ known from the previous-order solutions. However, at an arbitrary initial condition, the small perturbation can, in general, be a sum of all possible modes of the linearized system which leads to more complicated transformations, typically with the same lower-order results. In this section we briefly consider a more general, non-stationary scheme developed in the works of Keener and McLaughlin [11] and Gorshkov [4].

We again consider the general system (5.1), having at $\mu = 0$ a family of localized solutions, $\mathbf{u} = \mathbf{U}(\zeta, \mathbf{A})$ which tend to constants at $\zeta \to \pm\infty$. At $\mu \neq 0$ the solution in the vicinity of any given solitary wave is represented in the form

$$\mathbf{u}(t,\mathbf{r}) = \mathbf{U}(\zeta,\mathbf{A}) + \sum_n \mu^n \mathbf{u}^{(n)}(t,\mathbf{r}), \tag{5.17}$$

which differs from (5.3) in that the perturbations are allowed to have an arbitrary dependence on $\mathbf{r}$ and t. Substitution of this solution into (5.1) and equating the terms of the same order in μ results in the linear equations for perturbations, which are similar to (5.4):

$$G\mathbf{u}^{(n)} = \mathbf{H}^{(n)},$$
$$G = A^{(0)}\frac{\partial}{\partial t} + \mathbf{q}\mathbf{B}^{(0)}\frac{\partial}{\partial \mathbf{r}} + \left(\frac{\partial A^{(0)}}{\partial \mathbf{U}} - \mathbf{q}\frac{\partial \mathbf{B}^{(0)}}{\partial \mathbf{U}}\right)\frac{\partial \mathbf{U}}{\partial \zeta} + \frac{\partial C^{(0)}}{\partial \mathbf{U}}, \tag{5.18}$$

where the vector $\mathbf{H}^{(n)}$ contains functions defined in previous approximations; at $n = 1$ it has the same form as in (5.5). Note that here G is a partial-difference linear operator depending on $\mathbf{r}$ and t.

To construct the approximate solution, let us consider the initial-value problem for (5.18). As before, we assume that the basic solution $\mathbf{U}(\zeta, \mathbf{A})$ is stable with respect to initial perturbations. This means that the solution of the homogeneous part of (5.18), $G\mathbf{u} = 0$, is bounded for any finite initial perturbations. Thus, we focus on the behavior of the forced part of the solution which is generated by the function $\mathbf{H}^{(n)}$. We represent the general solution of (5.18) in terms of the complete set of eigenfunctions of the operator G:

$$\mathbf{u}^{(n)}(\mathbf{r},t) = \sum_d C_d^{(n)}(T)\mathbf{f}_d(\mathbf{r},t) + \int C_c^{(n)}(\mathbf{k},T)\mathbf{f}_c(\mathbf{k},\mathbf{r},t)d\mathbf{k}\,, \quad (5.19)$$

where $\mathbf{f}_d$ and $\mathbf{f}_c$ are eigenfunctions of the discrete and continuous parts of the spectrum, respectively, and $C_d(T)$ and $C_c(\mathbf{k}, T)$ are the corresponding coefficients to be determined. This expression includes both the summation over the discrete spectrum and integration over the continuous spectrum. Note that in general the continuous part can include both radiation and localized perturbations.

Substituting (5.19) into (5.18), we obtain equations for the coefficients C. For the system (5.18) which is represented in a normal form with respect to time derivatives (which in principle can always be done by a standard algebraic transform), equations for the coefficients are

$$\frac{dC_{d,c}(\mathbf{k})}{dT} = \int \mathbf{f}^*_{d,c}(\zeta,t)\mathbf{H}^{(n)}(\mathbf{r},t)d\zeta, \quad (5.20)$$

where $\mathbf{f}^*_{d,c}(\zeta,t)$ are the corresponding eigenfunctions of the discrete and continuous spectrum of the conjugated operator G^*.

Characteristic of such solutions is again a secular (i.e., of power-type, $\sim t^\alpha,\ \alpha > 0$) dependence on time of some of the coefficients $C_d(T)$ and $C_c(\mathbf{k}, T)$. Such conditions inevitably appear if some functions $\mathbf{f}$ in the right-hand side of (5.20) have a non-zero time average. For the functions of the discrete spectrum, it is indeed a problem which must be resolved (see below). However, secular divergence of the continuous-spectrum coefficients $C_c(\mathbf{k}, t)$ does not necessarily lead to an unbounded growth of the corresponding part of the solution of (5.20). This can be explained as follows. Eigenfunctions of continuous spectrum, $\mathbf{f}_c(\mathbf{k}, \mathbf{r}, t)$, are non-localized functions; they are not quadratically integrable on x. They are usually interpreted in terms of "incident" and "scattered" waves. The secular growth of coefficients $C_c(\mathbf{k}, T)$ corresponds to spreading of the domain occupied by the radiation from the solitary wave, rather than to the unbounded growth of the radiated field itself. On the contrary, the portion of (5.19) related to the eigenfunctions of the discrete spectrum remains localized in the vicinity of the basic solution $\mathbf{U}(\zeta, \mathbf{A})$; thus, increase of the coefficients $C_d(T)$ results in destruction of the basic solution and should be prevented.

Here again, among the localized eigenfunctions there are always those generated by variations of the basic solution $\mathbf{U}(\zeta, \mathbf{A})$ with respect to the parameters $\mathbf{A}$. In cases when the discrete part of the spectrum of eigenfunctions consist only of such functions, the solution (5.17) in the first approximation (up to the order of μ inclusive) can be represented as

$$\mathbf{u}(\mathbf{r},t)=\mathbf{U}(\zeta,\mathbf{A})+\mu t\sum_i \alpha_i \frac{\partial \mathbf{U}}{\partial A_i}+\mathbf{CS}\approx \mathbf{U}(\zeta,\mathbf{A}+\mu\boldsymbol{\alpha}t)+\mathbf{CS}, \quad (5.21)$$

where $\boldsymbol{\alpha}$ is a constant vector, and **CS** stands for a portion of the solution related to the continuous spectrum, which remains bounded. It is evident from this formula that the resonant (growing in the first order) portion of the perturbation causes the wave shift as a whole in the space of parameters $\mathbf{A}$. This conclusion provides a hint of how to suppress a secular divergence. One should treat the parameters A_i of the basic solution $\mathbf{U}(\zeta, \mathbf{A})$ as slowly varying functions of time, $A_i = A_i(T)$. This leads to a modification of the right-hand side of (5.18). For instance, in the first approximation, the vector $\mathbf{H}^{(1)}$ should be replaced with $\tilde{\mathbf{H}}^{(1)} = \mathbf{H}^{(1)} - \frac{\partial \mathbf{U}}{\partial \mathbf{A}}\frac{d\mathbf{A}}{dT}$. Now we can use the uncertainty in the choice of functions $A_i(T)$ to satisfy the following orthogonality conditions:

$$\int_{-\infty}^{\infty} \mathbf{f}_d^*(\zeta,t)\tilde{\mathbf{H}}^{(1)}(\mathbf{r},t)\,d\zeta = 0, \quad (5.22)$$

which prevent growth of coefficients of the discrete spectrum.

Equations (5.22) can be treated as differential equations which determine the slow dependence of parameters A_i on time. This dependence defines the evolution of the basic solution $\mathbf{U}(\zeta, \mathbf{A})$ under the action of perturbation. It should be emphasized that in the aforementioned case in which all eigenfunctions of the discrete spectrum are determined by the variations of $\mathbf{U}_{Ai}$, all secularities can be removed from the solution by means of orthogonality conditions (at least in the first approximation, the number of resonances is equal to the number of unknown functions A_i). In such cases the approximate solution of the basic set of PDEs reduces to the solution of ODEs of a finite order,

$$\frac{dA_i}{dT}\int \mathbf{f}_d^* \frac{\partial \mathbf{U}}{\partial A_i} d\mathbf{r} = \int \mathbf{f}_d^* \mathbf{H}^{(1)}(\mathbf{r}, t)\, d\mathbf{r}, \tag{5.23}$$

where $d = 1, 2, \ldots, m$.

As a result, the basic solution $\mathbf{U}$ with slowly varying parameters is determined. The small non-stationary corrections to this solution can be found from the expressions (5.19) which, after satisfying the orthogonality conditions (5.22), are defined by the functions of the continuous spectrum.

Note that the suppression of resonances described above corresponds to the quasi-stationary approximation discussed in the previous sections, but the quasi-stationary scheme is typically much simpler. Indeed, the analysis of the full expression (5.19) for perturbations is not a simple problem even in the rare cases when the continuous-spectrum eigenfunctions of operators G and G^* are known in an explicit form. In most cases it suffices to limit ourselves by considering the quasi-stationary (and thus potentially resonant) perturbations for which the previous scheme is certainly preferable. It is only when small non-stationary perturbations (included in the radiation fields) are important that it may be necessary to apply the non-stationary scheme described in this section.

For an equation close to fully integrable, the authors of [12] found an explicit Green function corresponding to both the discrete and continuous parts of the perturbation. This is possible for any N-soliton solution and even for their generalizations in the form of an N-periodic solution.

In [3] the variation principle was used in the framework of a method close to that described above in this section.

5.4. Inverse Scattering Perturbation Scheme for Solitons

In this section the class of localized waves is reduced to solitons understood as localized conservative waves which preserve (or exchange) their parameters after their interaction. Such waves are typically particular solutions of fully integrable equations. For systems close to integrable, the following approach (albeit leading to the results close to those obtained by the schemes described above) is applicable. It is based on the inverse scattering transform (IST) method. It is beyond the scope of this book to discuss in detail the exquisite inverse scattering theory for nonlinear PDEs; this theory is thoroughly discussed in the literature; see, for example, [1]. We reiterate that IST in its different forms sets up a correspondence between the initial nonlinear PDE and a linear equation containing a variable parameter (potential) depending on the sought solution $u(x,t)$ of the nonlinear equation, and a parameter λ defining the eigenvalues of the corresponding linear operator. Scattering of a plane harmonic wave from the potential is considered, and then an inverse problem, that of retrieval of the scattering potential from the scattering data (scattering coefficient and its phase), should be solved; as a result, the solution $u(x,t)$ is eventually found. In particular, if the scattering potential is non-reflecting and the scattering spectrum is discrete $(\lambda_1, \lambda_2, \ldots \lambda_N)$, the potential (and hence the solution) consists of N interacting solitons.

The highlights of this scheme are as follows [8–10, 14, 15]. As long as the exact N-soliton solution of the non-perturbed equation is known by using the IST, this method allows one to describe a slow evolution of solitons due to the action of perturbations. According to this scheme, the basic equation for a variable $u(x, t)$ (in scalar variables) is represented in an evolutional form

$$\frac{\partial u}{\partial t} = S[u] + \mu R[u], \tag{5.24}$$

where $S[u]$ and $R[u]$ are nonlinear differential operators, and it is supposed that at $\mu = 0$ the system (5.24) is integrable. An integrable equation can be represented via two linear operators, M and P, depending on u and known as the Lax pair:

$$\frac{\partial M}{\partial t} = [P, M] = PM - MP. \tag{5.25}$$

It can be shown that in the corresponding eigenvalue problem, $M\psi = \lambda\psi$, the eigenvalues λ do not depend on time. Thus, the problem can be solved for $t = 0$ and then extended to $t > 0$ by solving the equation $\psi_t(x,t) = P\psi(x,t)$ with an initial condition $\psi(x,0)$. Finally, this allows one to retrieve the "potential" u. For many typical integrable evolution equations, this is equivalent to considering a pair of linear equations for the function $\psi(x,t)$:

$$\frac{\partial \psi}{\partial x} = S(\lambda, u)\psi, \quad \frac{\partial \psi}{\partial t} = Q(\lambda, u)\psi. \tag{5.26}$$

Here the matrices S and Q consist of elements depending on the spectral parameter λ and having the solutions u of the basic evolution equation (5.24) (at $\mu = 0$) as parameters. Let us consider a particular case when both S and Q are 2×2 square matrices. In the spirit of IST, at the first stage the direct scattering problem is solved for the first equation (5.26) in which the variable t is treated as a parameter, since it is not explicitly present. As mentioned, the solution procedure is equivalent to finding the scattering coefficient $r(\lambda)$ (where $-\infty < \lambda < \infty$), together with eigenvalues λ_n and amplitude coefficients b_n for discrete-spectrum eigenfunctions. This scattering data can be expressed in terms of the so-called Yost coefficients a and b which relate two equivalent sets of eigenfunctions of continuous spectrum: $r(\lambda,t) = b(\lambda,t)a^{-1}(\lambda,t)$. The eigenvalues λ_n are defined by zeros of the function $a(\lambda, t)$ on the upper half-plane of the complex parameter λ, where the function a is analytical. Then the equations for a and b follow from the analysis of the asymptotic behavior of the continuous-spectrum functions at $|x| \to \infty$, i.e., from the scattering data. They are

$$\frac{\partial a(\lambda,t)}{\partial t} = 0, \quad \frac{\partial b(\lambda,t)}{\partial t} = i\Omega(\lambda) b(\lambda,t). \tag{5.27}$$

Here the function $\Omega(\lambda)$ coincides with the left-hand side of the dispersion relation $\omega(k) = 0$ for the linearized equation at $\mu = 0$. From the analytical properties of functions a and b in the upper half-plane of λ, one obtains equations for λ_n and b_n:

$$\frac{\partial \lambda_n}{\partial t} = 0, \quad \frac{\partial b_n(\lambda,t)}{\partial t} = i\Omega(\lambda_n) b_n(t). \tag{5.28}$$

From the scattering data known for all t, one can unambiguously restore $u(x,t)$ by solving the inverse scattering problem. Note that all known N– soliton solutions are associated with non-reflecting potentials, for which $r(\lambda,t) \equiv 0$. In these cases the constants λ_n define the amplitudes and velocities of solitons before and after interaction, whereas the functions $b_n(\lambda_n,t)$ describe their coordinates (phases). From the constancy of λ_n in (5.28) it directly follows that the number and parameters of the solitons are the same before and after the interaction ("elastic collision").

Let us return to the perturbed Eqs (5.24). At $\mu \neq 0$ (5.24) is still associated with the linear system (5.26), so that the direct scattering problem remains unchanged. However, temporal evolution of scattering data no longer obeys Eqs (5.27) and (5.28). One of the possible ways to describe this evolution is based on representation of scattering data as functionals F of the "scattering potentials" $u(x, t)$. By definition, the time derivatives of these functionals are represented as

$$\frac{dF}{dt} = \int_{-\infty}^{\infty} \frac{\delta F}{\delta u} \frac{\partial u}{\partial t} dx, \quad F = \{a, b, \lambda_n, b_n\}, \tag{5.29}$$

where the variational derivatives $\delta F/\delta u$ can be expressed in terms of eigenfunctions of the scattering problem. For (5.24) this relationship takes the form

$$\frac{dF}{dt} = \int_{-\infty}^{\infty} \frac{\delta F}{\delta u} S(u) dx + \mu \int_{-\infty}^{\infty} \frac{\delta F}{\delta u} R(u) dx, \tag{5.30}$$

where the first term in the right-hand side is evidently the same as for $\mu = 0$. As a result, evolution equations for the scattering data are perturbed with respect to those in (5.27):

$$a_t(\lambda,t) = \mu A(u,\lambda), \quad b_t(\lambda,t) = i\Omega(\lambda)b(\lambda,t) + \mu B(u,\lambda),$$
$$\lambda_{nt} = \mu\Lambda_n(u), \quad b_{nt} = i\Omega(\lambda_n)b_n(t) + \mu B_n(u). \tag{5.31}$$

Here A, B, Λ_n and B_n can be defined from the second term in (5.30), but only as functionals of the unknown solution $u(x,t)$. However, one can use the expansion of u in powers of μ similar to that in (5.3) and the corresponding expansions for the unknown parameters; for example, $A(u) = A(U) + \mu A^{(1)} + ...$, where $U(x, t)$ is again the basic solution known for $\mu = 0$ (it is typically an N– soliton solution). As a result, in the first approximation the functional derivatives are transformable to explicit functions of U depending on parameters λ and β, which closes the system (5.31). Their solution provides a slow time dependence of scattering data. Then, by restoring the scattering potential, one finds the perturbation $u^{(1)}(x,t)$. In the same manner, higher-order terms can be added to the system (5.31).

5.5. "Equivalence Principle"

It can be shown that, as expected, the results obtained in the framework of different perturbation schemes are essentially the same. In particular, the "direct" and "inverse" schemes (when the latter is applicable) yield identical equations for soliton parameters, in spite of all formal differences [16]. In many typical examples of perturbed equations, including KdV and other evolution equations to be considered further, the basic soliton $U(x, t)$ depends on one phase variable $\xi = x - V(\lambda_0)t - X_0$ and contains two parameters, λ_0 and X_0. The equations for these parameters, obtained from the inverse scheme in the first approximation, have the form

$$\frac{d\lambda_0}{dt} = \mu E(\lambda_0, X_0, t), \quad \frac{dX_0}{dt} = V(\lambda_0) + \mu D(\lambda_0, X_0, t). \tag{5.32}$$

When solving the same problem with the direct method, in the first approximation one obtains the corresponding equations which do not contain the last term (μD) which is present in the second equation (5.32). This difference is due to the different numbers of localized eigenfunctions used in the two schemes. Whereas the solution $U_{X0} \sim U_\xi$ is present in both schemes, in the quasi-stationary approach U_V is not an eigenfunction. This circumstance was mentioned in, e.g, [14]. However, the term with D is in fact a next-order (non-adiabatic) correction to the soliton velocity. In the direct scheme a similar correction ($\sim U_V$) appears in the expression for the perturbation $u^{(1)}$. Moreover, the use of the term μD to find a correction to the phase variable is consistent only together with terms of the order of μ^2 in the first equation (5.32); this fact is well known in oscillation theory [2]. Thus, in general, the two schemes satisfy the "equivalence principle": they essentially produce the same results.

5.6. Concluding Remarks

In this chapter, general schemes of the asymptotic perturbation theory of localized waves were presented. It should be stressed that specific cases of asymptotic dynamics of solitons have been discussed since the 1970s in a number of works, including [6, 7, 13], to name a few. Other examples, as well as various physically sound applications, will be given in subsequent chapters. In these examples the direct method will be used mainly in its quasi-stationary form because of its relative simplicity and universality: as mentioned, the corresponding scheme can be constructed in a similar way for integrable and non-integrable equations; besides, this method has much in common with that used above for quasi-periodic waves.

References

1. Ablowitz, M. J. and Segur, H. (1981). *Solitons and the Inverse Scattering Transform.* SIAM, Philadelphia.
2. Bogoluybov, N. N. and Mitropolsky, Y. A. (1961). *Asymptotic methods in oscillation theory.* Gordon and Breach, New York.
3. Bondeson, A., Lisak, M., and Anderson D. (1979). Soliton perturbations. A variational principle for the soliton parameters, *Physica Scripta*, v. 20, pp. 479–485.
4. Gorshkov, K. A. (2007). *Perturbation Theory in Soliton Dynamics* (Doctor of Science Thesis, Institute of Applied Physics, Nizhni Novgorod, Russia) [in Russian].
5. Gorshkov, K. A. and Ostrovsky, L. A. (1981). Interactions of solitons in non-integrable systems: Direct perturbation method and applications, *Physica D*, v. 3, 428–438.
6. Gorshkov, K. A., Ostrovsky, .L. A., and Pelinovsky, E. N. (1974). Some Problems of Asymptotic Theory of Nonlinear-Waves, *Proceedings of the IEEE*, v. 62, pp.1511–1517.
7. Grimshaw R. (1979). Slowly varying solitary waves. I. Korteweg–de Vries equation, and II. Nonlinear Schrödinger equation, *Proceedings of the Royal Society*, v. 368A, 359–375 and 377–388.
8. Karpman, V. I. and Maslov, E. M. (1977). Perturbation theory for solitons, *Soviet Physics JETP*, v. 73, pp. 281–291.
9. Karpman, V. I (1979). Soliton evolution in the presence of perturbations, *Physica Scripta,* v. 20, pp. 462–468.
10. Kaup, D. J. and Newell, A. C. (1978). Solitons as particle, as oscillators, and in slowly changing media: A singular perturbation theory. *Proceedings of the Royal Society*, v. 301, pp. 413–446.
11. Keener J. R. and McLaughlin, D.W. (1977). Solitons under perturbation, *Physical Review A*, v. 16, pp. 777–790.
12. Keener J. R. and McLaughlin, D.W. (1977). A Green's function for a linear equation associated with solitons, *Journal of Mathematics and Physics*, v. 18, pp. 2008–2013.
13. Ko, K. and Kuehl, H. H. (1978). Korteweg–de Vries soliton in a slowly varying medium. *Physical Review Letters*, v. 40, pp. 233–236.
14. Kodama, Y. and Ablowitz, M. (1981). Perturbations of solitons and solitary waves. *Studies in Applied Mathematics*, v. 64, pp. 225–245.
15. Malomed, B. A. and Kivshar, Y. S. (1989). Dynamics of solitons in nearly integrable systems, *Review of Mathematics and Physics*, v. 61, pp. 763–915.
16. Ostrovsky, L. A. and Gorshkov, K.A. (2000). Perturbation theories for nonlinear waves, in *Nonlinear Science at the Dawn of the 21st Century*, eds. P. Christiansen, M. Soerensen, and A. Scott. Springer, Berlin, pp. 47–66.

Chapter 6

Perturbed Solitons

> A man should look for what is, and not for what he thinks should be.
>
> Albert Einstein

This chapter deals with solitons, the solitary, conservative nonlinear waves. It demonstrates how to apply the general perturbation theory developed in the previous chapter to specific wave models. As already mentioned, since the 1960s the theory of solitons has been established as one of the most important chapters of the modern theory of nonlinear waves. The examples considered below have an independent practical interest as well, since they belong to rather typical classes of waves having applications in different areas of science.

It should be emphasized that here we define a soliton as a conservative solitary wave, not necessarily remaining unchanged upon interaction with another such wave, similar to an elastic collision of two or more material particles. The latter feature, which is characteristic of integrable nonlinear equations, was used as an early general definition of a soliton [28]. However, in many cases, including most non-integrable systems, interaction of solitary waves produces a small radiation, whereas the waves still preserve their entity as particles. We find this a sufficient reason to include such waves in the soliton family. Now this extended definition of a soliton seems to be well accepted in the literature.

6.1. Perturbed KdV Equation

6.1.1. *Equation for soliton amplitude*

Consider first the Korteweg–de Vries (KdV) equation, for which periodic solutions have already been constructed in Chapter 4. In various physical situations the KdV equation can be perturbed by small factors such as dissipation, inhomogeneity, geometrical divergence, focusing, etc. In general, these additions make the equation non-integrable but, as already mentioned, the integrability is not a crucial factor for the direct perturbation theory. Examples of slowly varying solitons in different physical realizations were considered in many publications beginning from the early papers [8, 9, 16, 31].

Consider the KdV equation (4.46) with a small additional term:

$$u_t + uu_x + u_{xxx} = \mu R(u). \tag{6.1}$$

It is assumed that in the main order, $R(u = U)$ vanishes together with U at $x \to \pm\infty$; here U is the solitary solution of (6.1) at $R = 0$.

In Chapter 4, the expression (4.48) describes stationary periodic solutions of the non-perturbed equation (4.46). When the nonlinearity parameter s in the periodic solution (4.48) tends to unity, one obtains a limiting solitary solution, a soliton, represented by the closed separatrix in the phase plane, Fig. 4.4. It has the form:

$$u = U(\zeta) = A\,\mathrm{sech}^2(\zeta / \Delta),\ V = A/3,\ \Delta = \sqrt{12/A}, \tag{6.2}$$

where $\zeta = x - Vt$ and V is the soliton velocity (for convenience, we slightly changed the form of the argument ζ). We remind readers that in the physical coordinates the KdV equation has the form (4.45), and the soliton velocity is $c_0 + V$, where c_0 is the velocity of long linear waves. The parameters A and Δ are the soliton amplitude and its characteristic width (Fig. 6.1).

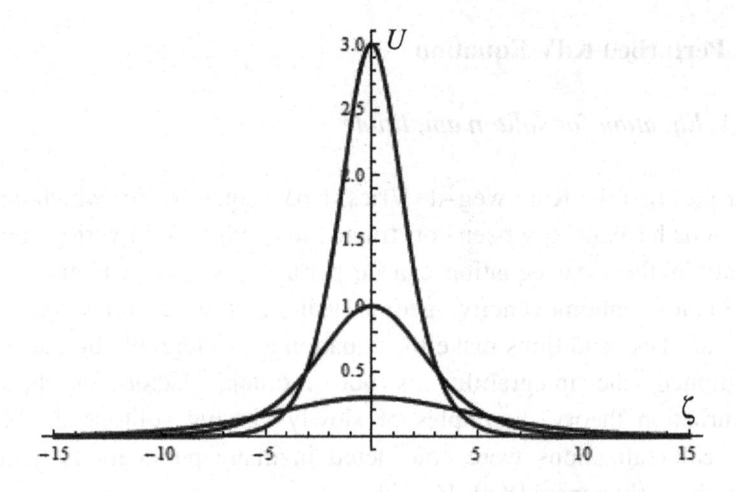

Figure 6.1. Solitons in the KdV equation (6.1) with $R = 0$, having amplitudes $A = 0.3$, 1.0, and 3.0. At the level of $U \approx 0.42A$, the soliton width is equal to 2Δ.

According to (6.2), the parameters A, V, and Δ characterizing a soliton are related, so that only one of them is independent.

Let us now return to the full Eq. (6.1) with $R(u) \neq 0$ and apply the direct perturbation theory described in Chapter 5. We seek an approximate solution of (6.1) in the form

$$u(x,t) = A(T)\operatorname{sech}^2\left(\sqrt{\frac{A(T)}{12}}\zeta\right) + \sum_{n=1}^{J} \mu^n u^{(n)}(\zeta, T), \tag{6.3}$$

where again $T = \mu t$ is a "slow" time and $\zeta = x - \int V(T)dt$.

Substituting this solution into (6.1), we obtain in each approximation the linear equations (5.4) of Chapter 5, namely:

$$Gu^{(n)} = \frac{d}{d\zeta}\left[-V + U(\zeta) + \frac{d^2}{d\zeta^2}\right]u^{(n)} = H^{(n)}, \tag{6.4}$$

with $U = U(\zeta, T)$ defined by (6.2).

The conjugate counterpart of the operator G is

$$G^* = (V - U(\zeta))\frac{d}{d\zeta} - \frac{d^3}{d\zeta^3}. \tag{6.5}$$

Fundamental solutions of the homogeneous equation $G\psi = 0$ can be presented in terms of the solitary solution U:

$$\psi_1 = U_\zeta,\ \psi_2 = 1 - U_V = U_\zeta \int_0^\zeta \frac{U}{U_\zeta^2} d\zeta',\ \psi_3 = U_\zeta \int_0^\zeta \frac{d\zeta'}{U_\zeta^2}. \tag{6.6}$$

The corresponding fundamental solutions of the conjugated equation $G^*\psi = 0$ are

$$\psi_1^* = U(\zeta),\ \psi_2^* = 1,\ \psi_3^* = \int_0^\zeta U_\zeta \left(\int_0^{\zeta'} \frac{d\zeta''}{U_\zeta^2} \right) d\zeta'. \tag{6.7}$$

According to the general scheme of Chapter 5, the solution of Eq. (6.4) in each order n has the form (5.6). After using the functions (6.6) and (6.7), this solution can be written as

$$\begin{aligned} u^{(n)} &= U_\zeta \left(C_1^{(n)} + \int_0^\zeta d\zeta' H^{(n)} \left[\int_0^{\zeta'} d\zeta'' \left(U_\zeta \int_0^{\zeta''} d\zeta''' / U_\zeta^2 \right) \right] \right) \\ &+ \left(U_\zeta \int_0^\zeta d\zeta' U / U_\zeta^2) \right) \left(C_2^{(n)} + \int_0^\zeta H^{(n)} d\zeta' \right) \\ &+ \left(U_\zeta \int_0^\zeta d\zeta' / U_\zeta^2 \right) \left(C_3^{(n)} - \int_0^\zeta d\zeta' U H^{(n)} \right). \end{aligned} \tag{6.8}$$

It follows from (6.8) and the asymptotic values of $U(\zeta)$ at $\zeta \to \pm\infty$, that for the solution $u^{(n)}$ to remain bounded, two orthogonality conditions corresponding to (5.10) and (5.11) of Chapter 5 must be fulfilled:

$$\int_{-\infty}^{\infty} U H^{(n)} d\zeta = 0, \quad \lim_{\zeta \to \pm\infty} H^{(n)} = 0. \tag{6.9}$$

The first-order function $H^{(1)}$ has a simple form:

$$H^{(1)} = R(U) - U_T. \tag{6.10}$$

As assumed, $R(U)$ vanishes together with U at $\zeta \to \pm\infty$; this implies that $H^{(1)}$ also vanishes when $\zeta \to \pm\infty$. Hence, the second algebraic condition of (6.9) is met identically. After using the zero-order solution (6.2) for U, the first integral condition (6.9) results in the equation for soliton amplitude:

$$\frac{dA}{dT} = \frac{4}{3}\frac{\int_{-\infty}^{\infty} \varphi(\theta)R[\varphi(\theta)]d\theta}{\int_{-\infty}^{\infty} \varphi^2(\theta)d\theta} = \int_{-\infty}^{\infty} \operatorname{sech}^2\theta \cdot R\left[\operatorname{sech}^2\theta\right]d\theta, \quad (6.11)$$

where $\theta = \zeta/\Delta$ and $\varphi(\theta) = U(\theta)/A = \operatorname{sech}^2\theta$.

Equation (6.11) describing the variation of soliton parameters in the first approximation can also be obtained from the basic equation (6.1) as a "balance equation". After multiplying (6.1) by $u(x,t)$, integrating over the entire x-axis and taking into account that the solution vanishes at infinity, we arrive at the following equation:

$$\frac{1}{2}\frac{d}{dT}\int_{-\infty}^{\infty} u^2 dx = \int_{-\infty}^{\infty} uRdx. \quad (6.12)$$

Substituting the zero-order solution (6.2) into (6.12), we obtain the same Eq. (6.11) for the soliton amplitude.

Note that for the periodic functions considered in Chapters 3 and 4, Eq. (6.12) is also true if the integrals are taken over the period rather than infinite limits.

The quantity

$$P = \frac{1}{2}\int_{-\infty}^{\infty} u^2 dx \quad (6.13)$$

can be treated as total momentum of the soliton. This interpretation follows from the Lagrangian representation of the KdV equation [e.g., 24]. On the other hand, for KdV in its "physical" form (4.45), P can be considered as wave energy if the additions from small nonlinearity and dispersion are neglected.

6.1.2. *KdV equation with dissipation*

Let us specify some forms of the function $R(U)$ and consider its impact on the variation of soliton parameters (amplitude, velocity, width). Note that if R consists of several additive terms, the first-order equation for soliton amplitude will also be the sum of the corresponding terms.

Consider first how solitons are attenuated due to different kinds of energy losses. Let Eq. (6.1) have the form

$$u_t + uu_x + u_{xxx} = -\mu(qu + m\,|\,u\,|\,u - gu_{xx}). \tag{6.14}$$

The last term on the right-hand side of (6.14) is responsible for the "Reynolds viscosity"; in the absence of the other two terms, (6.14) corresponds to the Korteweg–de Vries–Burgers (KdVB) equation describing waves in a dispersive medium (such as acoustic and magnetoacoustic waves in plasma) with viscous dissipation. The first and third terms are also used for the description of electromagnetic waves in nonlinear transmission lines. The second nonlinear term was suggested as a model for friction (a version of the Chezy law [3]); it is often used to describe dissipation of water waves due to bottom friction [22]. Various examples of such equations in physics and hydrodynamics can be found in the books [25, 31] and in the numerous publications referred to in these books.

Now Eq. (6.11) for the soliton amplitude reduces to

$$\frac{dA}{dT} = -\frac{4A}{3\Delta^2}\frac{\int_{-\infty}^{\infty}\left[q\varphi^2 + mA\varphi^3 + g\varphi_\theta^2/\Delta^2\right]d\theta}{\int_{-\infty}^{\infty}\varphi^2 d\theta} = \tag{6.15}$$

$$-\left(\frac{4}{3}qA + \frac{4gA^2}{45} + \frac{16mA^2}{15}\right),$$

having the solution

$$A = \frac{A_0 e^{-4qT/3}}{1 + \frac{pA_0}{15q}\left(1 - e^{-4qT/3}\right)}, \; p = g + 12m. \tag{6.16}$$

Here $A_0 = A(T = 0)$. As soon as the variation $A(T)$ is known, soliton width $\Delta(T)$ and its velocity $V(T)$ are defined by (6.2).

The following particular cases of (6.16) are worth mentioning. If $p = 0$, soliton amplitude decays exponentially: $A = A_0 e^{-4qT/3}$. Note that this decay rate of a soliton, $4q/3$, is larger than that of a linear non-dispersive wave, q. The physical explanation of the faster decay of a soliton is in the dependence of its width on amplitude: in the course of damping, the soliton broadens (the width Δ increases), so that at the same rate of attenuation of soliton energy, which is proportional to $A^2\Delta \sim e^{-2qT}$, the amplitude decreases faster.

If $q = 0$, (6.16) becomes

$$A = \frac{A_0}{1 + pA_0 T / 45}. \tag{6.17}$$

In this case the damping is non-exponential. It is remarkable that at a large time interval, when $A_0 T >> 45/p$, we have $A \approx 45 / T$, i.e., the amplitude no longer depends on its initial quantity. A similar feature is known for weak shock waves in nonlinear gas dynamics and acoustics [4, 18]. Note also that the nonlinear Chezy term with m makes a soliton attenuate in the same way as the linear viscous term with g. However, this similarity is limited to the specific relation between soliton amplitude and width; in general these two types of dissipation can produce quite different solutions, especially if the solution changes its sign.

6.1.3. *Radiation from the soliton*

According to Chapter 5, if the orthogonality conditions are met, the additions to the localized solution $U(\zeta, A)$ are bounded at all ζ. Still, a possible non-localized perturbation component $u^{(n)}(T, X)$ governed by the second equation (6.9) deserves a special consideration. Thus, consider the behavior of perturbations at $\zeta \to \pm\infty$. According to (6.2), the soliton vanishes as $\exp(-A^{1/2}|\zeta|/3) = \exp(-V^{1/2}|\zeta|)$, and at large $|\zeta|$ the perturbation $u^{(1)}$ can exceed the basic function U. Suppose that the function $R(U)$ in Eq. (6.1) is a polynomial of U and its derivatives. Then the function $H^{(1)} = R(U) - U_T$ vanishes at $\zeta \to \pm\infty$ not slower than $\zeta\exp(-V^{1/2}|\zeta|)$; note that the factor ζ appears due to the term U_T.

It can be seen that as long as the conditions (6.9) are met, the non-localized (i.e. non-vanishing at $\zeta \to \infty$ or $-\infty$) component of perturbations $u^{(n)}$ is only due to the second term in the solution (6.8), the one that is proportional to the non-vanishing function $\psi_2(\zeta)$. The expression for this small, non-localized component of $u^{(1)}$ can be obtained directly from (6.4). Since at $|\zeta| >> \Delta$ the function U is exponentially small, in this external region Eq. (6.4) at $n=1$ has the form $Gu^{(1)} \approx -V\partial u^{(1)}/\partial\zeta = H^{(1)}$. Integrating this equation, one obtains

$$u^{(1)}(\zeta,T,X) \approx C_2^{(1)}(T,X) + \frac{1}{V}\int_0^{\zeta}(R(U) - U_T)\,d\xi' + O\left(\zeta^2 e^{-|\zeta|/\Delta}\right). \quad (6.18)$$

Now take into account that at $|\zeta| >> \Delta$ the integral of U_T has the form

$$\int_0^{\zeta} U_T d\zeta' = \frac{d}{dT}\int_0^{\zeta} U d\zeta' \approx 6\frac{d}{dT}\sqrt{V}\int_0^{\theta}\cosh^{-2}(\theta')d\theta'$$
$$= \frac{3}{\sqrt{V}}\frac{dV}{dT}\tanh(\zeta/\Delta) = \frac{\sqrt{3}}{\sqrt{A}}\frac{dA}{dT}\tanh(\zeta/\Delta), \quad |\zeta| >> \Delta. \quad (6.19)$$

Then, using the expression (6.2) with $A = 3V$, one can reduce (6.18) for large ζ to the following form:

$$u^{(1)}(|\zeta| >> \Delta, T, X) = C_2^{(1)}(T,X)$$
$$+\frac{1}{V^{3/2}}\int_0^{\zeta/\Delta} R(U)\left[\operatorname{sech}^2(\theta) - 2\right]d\theta. \quad (6.20)$$

At large distances this perturbation remains a slowly varying function of T and X which can be called radiation. The unknown function $C_2^{(1)}$ is defined by the algebraic condition (the second equation (6.9)) in the second approximation, $n = 2$; as mentioned, (6.9) is automatically met in the first approximation. If R is again a polynomial of u and its derivatives, then the part of $H^{(2)}$ that is non-vanishing at large distances is $-\partial u_{\pm}^{(1)} / \partial T + \delta u_{\pm}^{(1)}$, where δ is a constant. It is non-zero only if $R(u)$ contains a term proportional to u; the derivatives of u can be neglected in this order. Here the subscripts $\pm$ correspond to $\zeta \to \pm\infty$. Then from the algebraic condition it follows that

$$\partial u_{\pm}^{(1)} / \partial T - \delta u_{\pm}^{(1)} = 0, \qquad (6.21)$$

with the solution

$$u_{\pm}^{(1)} = e^{\delta T} F_{\pm}(X). \qquad (6.22)$$

The functions $F_{\pm}$ are defined by the initial condition, $u = u_{\pm}^{(1)}(X,0)$. Then one should match the functions $u_{\pm}^{(1)}$ on the trajectory of the soliton, $X = X_s(T)$, namely,

$$\left[u^{(1)}\right] = (u_{+}^{(1)} - u_{-}^{(1)})\Big|_{X=X_s} = \frac{1}{V^{3/2}} \int_{-\infty}^{\infty} d\theta R(U)\left[\mathrm{sech}^2(\theta) - 2\right]. \qquad (6.23)$$

The radiation field magnitude can be found from here. If, in particular, the initial condition defines a non-perturbed soliton, i.e., $u_{\pm}^{(1)}(X,0) = 0$, we have $u_{+}^{(1)} \equiv 0$ (no radiation ahead of the soliton), and the radiation field behind the soliton has the form of a slowly varying "shelf", decreasing in time and expanding into the region $0 < X < X_s$:

$$u_{-}^{(1)}(X,T) = u_{-}^{(1)}(T = T(X_s))\exp(-\delta T(X_s)) \qquad (6.24)$$

(see Fig. 6.2). Here $T(X_s)$ is the time when the soliton center is located at $X = X_s$, i.e., the function inverse to $X_s(T) = \int_0^T V(T')dT'$.

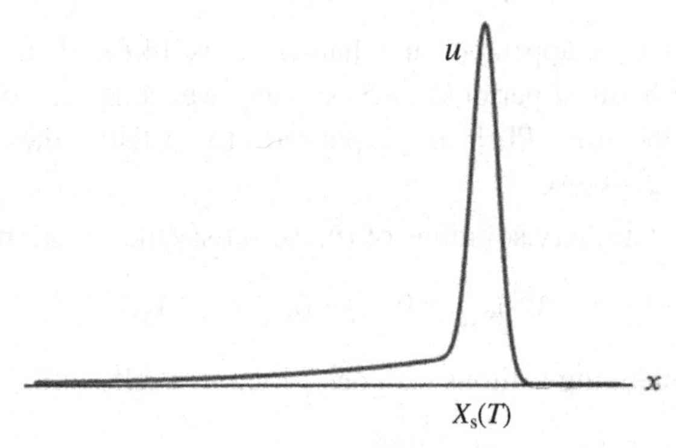

Figure 6.2. Schematic plot of a soliton with a radiated tail.

From comparison of (6.20) and (6.23) it follows that in (6.20) at the soliton trajectory,

$$C_2^{(1)}(X = X_s(T), T) = \frac{1}{2}u_-^{(1)}. \tag{6.25}$$

On the scale of the slow variable X, the solution (6.24) has a "jump" at $X = X_s$. In reality the "jump" is a transition area having a width of the order of the soliton width Δ. To describe this transition, it is necessary to take into account the non-steady radiation corresponding to the solution of the homogeneous equation $Gu = 0$, generated at the initial moment. A more detailed analysis [7] shows that at $T > 0$ the radiation field in the vicinity of the soliton center has the form of the Airy function which broadens and oscillates behind the soliton. In [8] it was shown that in the case or viscous losses when in (6.14), $q = m = 0$, $g \neq 0$, the formation of a "shelf" behind a soliton eventually transforms it to a triangle-shaped impulse attenuating according to the asymptotics of Burgers' equation which will be addressed in Chapter 8. In the same work the main stages of the process described above were experimentally observed in an electric line.

6.2. Nonlinear Klein–Gordon Equation

Consider now the perturbed Klein–Gordon (KG) equation:

$$u_{tt} - u_{xx} + P(u) = \mu R[u]. \tag{6.26}$$

A similar equation appeared in Chapter 4 as (4.8), then (4.40), in connection with quasi-periodic waves; here we consider solitons. As above, the operator $R[u]$ is supposed to satisfy the condition $R[U] \to 0$ at $\zeta \to \pm\infty$.

At $R = 0$, stationary solutions of (6.26) satisfy the equation

$$(1-V^2)u_{\zeta\zeta} + P'(u) = 0, \ \zeta = x - Vt. \tag{6.27}$$

Its solutions, including solitons, can be written in a self-similar form:

$$u^{(0)} = U\left(\frac{\zeta}{\sqrt{1-V^2}}\right). \tag{6.28}$$

Applying to (6.26) the same perturbation scheme as before, we obtain

$$Gu^{(n)} = \left[(1-V^2)\frac{d^2}{d\zeta^2} - P'(U)\right]u^{(n)} = H^{(n)}. \tag{6.29}$$

This example is simplified by the fact that the second-order operator G is self-adjoint. One of the two fundamental solutions, y_1, of equation $Gy = 0$, vanishing at the infinity, is known immediately: $y_1 = U_\zeta$; the other is $y_2 = U_\zeta \int_0^\zeta d\zeta' / U_\zeta^2$. Now the solution of (6.29) in each order can be written in quadratures:

$$u^{(n)} = y_1\left(C_1^{(n)} + \int_0^\zeta y_2 H^{(n)} d\zeta'\right) + y_2\left(C_2^{(n)} + \int_0^\zeta y_1 H^{(n)} d\zeta'\right). \tag{6.30}$$

To keep these perturbations bounded, it is necessary and sufficient to meet the orthogonality condition in each order of μ:

$$\int_{-\infty}^{\infty} U_\zeta H^{(n)} d\zeta = 0. \tag{6.31}$$

In the first approximation, $n = 1$,

$$H^{(1)} = U_\zeta \frac{dV(T)}{dT} + 2VU_{\zeta T} + R[U], \tag{6.32}$$

and, according to (6.31), the soliton velocity satisfies the equation

$$\frac{d}{dT}\left(\frac{V}{\sqrt{1-V^2}}\right) = -\frac{1}{\langle U_z^2 \rangle} \int_{-\infty}^{\infty} dz U_z(z) R[U(z)]; \; z = \frac{\zeta}{\sqrt{1-V^2}}. \tag{6.33}$$

As long as this equation is satisfied, both the first and second terms in (6.30) tend to zero at $\zeta \to \pm\infty$ as $\zeta^2 \exp(-\lambda|\zeta|)$ and $\zeta \exp(-\lambda|\zeta|)$ respectively, where $\lambda > 0$ corresponds to the soliton's asymptotic behavior far from its center. Hence, in this case all perturbations are localized near the soliton center, and radiation (at least, the long-wave type) is absent.

To be more specific, let us consider the KG equation in the form of (4.40):

$$u_{tt} - u_{xx} + u - u^3 = -\mu\nu u_t. \tag{6.34}$$

Periodic and quasi-periodic solutions of a similar equation was considered in Chapter 4. At $\mu = 0$, Eq. (6.27) for stationary waves has the form

$$\left(V^2 - 1\right) U_{\zeta\zeta} + U - U^3 = 0. \tag{6.35}$$

Consider the case of "slow" waves, $V^2 < 1$. In this case there exists an one-parameter family of solitons:

$$U = \sqrt{2}\,\mathrm{sech}\left(\zeta\sqrt{\frac{2}{1-V^2}}\right). \tag{6.36}$$

Here, unlike the KdV case, the soliton amplitude $A = \sqrt{2}$ is not a free parameter, whereas the soliton width in the physical variables depends on its velocity. If V is close to 1, which is the velocity of a linear wave in the high-frequency limit, the soliton is very narrow. If $V \to 0$, such an immovable soliton has a limiting shape with maximal energy. Figure 6.3 shows the shapes of these solitons at different velocities.

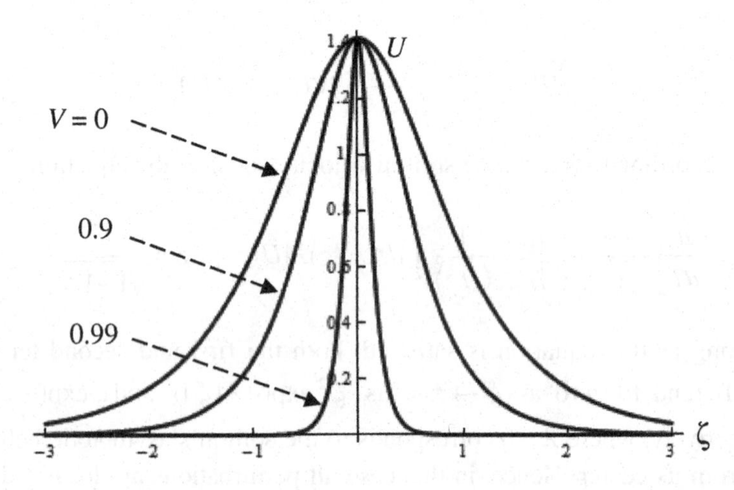

Figure 6.3. Profiles of the soliton (6.36) at different velocities V marked on the plot.

Substitution of the soliton (6.36) into (6.33), where $R = \nu\, U_t$ yields

$$\frac{d}{dT}\left(\frac{V}{\sqrt{1-V^2}}\right) = \nu \frac{V}{\sqrt{1-V^2}},$$
$$\text{so that } V = \frac{V_0 e^{\nu T/2}}{\sqrt{1+V_0^2\left(e^{\nu T}-1\right)}}, \tag{6.37}$$

where $V_0 = V(T = 0)$. Figure 6.4 shows the variation of soliton velocity (which defines, its width and energy) with time, at different initial conditions. Due to dissipation, the soliton velocity tends to 1 so that, according to (6.36) and Fig. 6.3, its energy tends to zero: in this approximation the soliton "disappears" by becoming infinitely narrow.

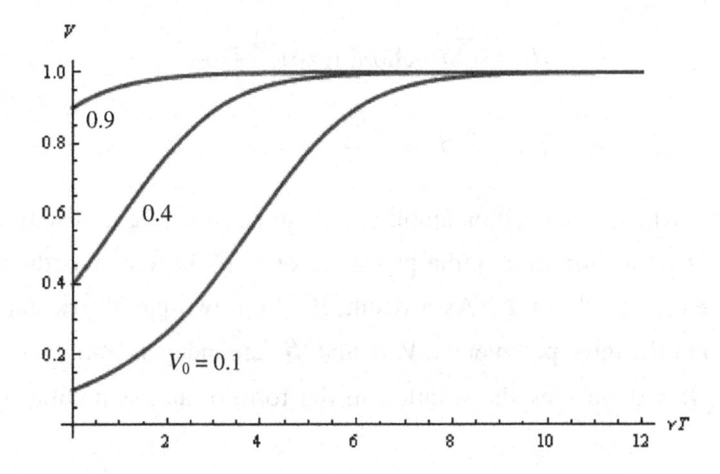

Figure 6.4. Increase of the soliton velocity V with time due to dissipation for three initial values V_0 marked at the curves.

6.3. Nonlinear Schrödinger Equation

The Nonlinear Schrödinger Equation (NSE) in the form (1.45) was obtained in Chapter 1 for a particular example; as mentioned, it is characteristic of a broad class of modulated waves in dispersive media with small cubic nonlinearity. Here we consider a perturbed NSE in the form

$$iA_t + A_{xx} + A|A|^2 - n(X,T)A = \mu R[A]. \qquad (6.38)$$

Here n is a given, slowly varying parameter describing, for example, a small variable addition to the effective refraction index in optical fibers [17, 21].

Note that in physical problems, A is commonly a slowly varying complex amplitude of a quasi-harmonic wave, and in Chapter 1 it was defined as a function of slow variables T and X. However, now the term in the right-hand part of (6.38) is supposed to be small as compared with each term in the left-hand part. Since the carrier wave is not present in (6.38), we use fast variables t and x for A in the soliton considered below, and slow variables T and X for its slow variations.

At $R = 0$ this equation has a family of solitary solutions ("envelope solitons"):

$$A^{(0)} = U = \sqrt{2}b\,\mathrm{sech}(b\zeta)\exp i\left(\frac{V\zeta}{2}+\varphi\right),$$
$$\varphi = \beta t,\ b^2 = \beta - n - \frac{V^2}{2},\ \zeta = x - Vt. \tag{6.39}$$

Note that whereas the soliton amplitude modulus, $\mathrm{sech}(b\zeta)$, is stationary (i.e., it is a function of ζ), the phase factor in (6.39) contains the term with t separately from ζ. As a result, this is a two-parametric family: any two of the three parameters, V, b, and β , are independent.

At $R \neq 0$ we seek the solution in the form of an asymptotic series [7]:

$$A = \left[U(\zeta,T,X) + \sum_n \mu^n u^{(n)}(\zeta,T,X)\right]\cdot \exp i\left(\frac{V\zeta}{2}+\varphi\right). \tag{6.40}$$

Substituting this into (6.38), we have in each approximation

$$Gu^{(n)} = H^{(n)},$$
$$G^{(n)} = \left(\frac{d^2}{d\zeta^2} - b^2 - n\right)u^{(n)},$$
$$H^{(n)} = -iu_T^{(n-1)} - iVu_X^{(n-1)} + \frac{V_T}{2}\zeta u^{(n-1)}$$
$$-2u_{\zeta X}^{(n-1)} - u_{XX}^{(n-2)} + R\left[u^{(n-1)}\right]\cdot\exp\left[-i\left(\frac{V\zeta}{2}+\varphi\right)\right]. \tag{6.41}$$

This is equivalent to the following two equations for real and imaginary parts of $u^{(n)}$:

$$\left(\frac{d^2}{d\zeta^2} - b^2 - 3n\right)\mathrm{Re}\,u^{(n)} = \mathrm{Re}\,H^{(n)}$$
$$\left(\frac{d^2}{d\zeta^2} - b^2 - n\right)\mathrm{Im}\,u^{(n)} = \mathrm{Im}\,H^{(n)}. \tag{6.42}$$

Two fundamental solutions of the homogeneous equation $Gu^{(n)} = 0$ vanishing at $\zeta \to \pm\infty$ can be found by variation of the soliton (6.39) with respect to ζ and φ; they are proportional to U_ζ and U, respectively. The remaining pair of fundamental solutions, y_r and y_i, that increase at $\zeta \to \pm\infty$, can be found from these two as described in the theory of differential equations (their specific form does not affect the orthogonality conditions (6.44)). As a result, the solutions of (6.42) can be expressed in quadratures:

$$\mathrm{Re}\, u^{(n)} = U_\zeta \left(C_{r1}^{(n)} + \int_0^\zeta y_r \,\mathrm{Re}\, H^{(n)} d\zeta' \right) + y_r \left(C_{r2}^{(n)} + \int_0^\zeta U_\zeta \,\mathrm{Re}\, H^{(n)} d\zeta' \right),$$

$$\mathrm{Im}\, u^{(n)} = U \left(C_{i1}^{(n)} + \int_0^\zeta y_i \,\mathrm{Im}\, H^{(n)} d\zeta' \right) + y_i \left(C_{i2}^{(n)} + \int_0^\zeta U \,\mathrm{Im}\, H^{(n)} d\zeta' \right). \tag{6.43}$$

To keep these functions limited, two necessary orthogonality conditions must be met:

$$\int_{-\infty}^{\infty} U_\zeta \,\mathrm{Re}\, H^{(n)} d\zeta = 0, \quad \int_{-\infty}^{\infty} U \,\mathrm{Im}\, H^{(n)} d\zeta = 0. \tag{6.44}$$

In general these equations can be insufficient since, as mentioned, U can depend on two fast variables, ζ and t. Thus, we impose an additional constraint by supposing that $R[U] = R_\alpha R_\beta$, where R_α depends on the modulus of U (and, possibly, slow variables), and R_β is a linear functional of U and its derivatives. This form is relevant to many physical applications. In this case R can be represented in the form

$$R[U] = R_1(|U|, X, T) \exp\left[i \left(\frac{V}{2}\zeta + \varphi \right) \right], \tag{6.45}$$

As a result, in the first approximation the conditions (6.44) yield the equations for the velocity V, and the amplitude (or inverse width) b of a soliton:

$$\frac{db}{dT} = \frac{b}{\sqrt{2}} \int_{-\infty}^{+\infty} \cosh^{-1} b\zeta \, \mathrm{Im}\, R[U] d\zeta,$$
$$\frac{dV}{dT} = -2\frac{\partial n}{\partial X} - \sqrt{2}b \int_{-\infty}^{+\infty} \frac{\sinh b\zeta}{\cosh^2 b\zeta} \mathrm{Re}\, R[U] d\zeta. \tag{6.46}$$

These equations define the functions b and V at the soliton trajectory, i.e., at $X = X_s(T) = \int_0^T V(T')dT'$. It can be shown that as long as the conditions (6.44) are met, all perturbations $u^{(n)}$ are not only limited but also localized. In particular, the first-order perturbations $u^{(n)}$ decrease at $\zeta \to \pm\infty$ as $\zeta^2 \exp(-\lambda|\zeta|)$ and $\zeta \exp(-\lambda|\zeta|)$ (they refer to the first and second terms in (6.43), respectively). Consequently, as in the KG case, there is no radiation "tail" moving with the soliton (albeit the free field can exist if the initial condition differs from the soliton).

Along with [7], the equations similar to (6.46) are also given in [17] with reference to [14].

If $R = 0$ but $n \neq 0$ (propagation in an inhomogeneous medium), Eqs (6.46) are reduced to

$$b = \text{const}, \quad \frac{dV}{dT} = -2\frac{dn}{dX}. \tag{6.47}$$

In particular, for a linear profile of $n(X) \propto X$, the soliton is uniformly accelerated, $V \propto T$, which coincides with the exact solution for this case. This provides evidence for the effectiveness of the approximate theory discussed here.

Another example considered in [17] refers to the case $n = 0$ and $R = -igA$, which may describe, e.g., losses in optical fibers. In this case, from the first equation (6.46), it follows that the amplitude parameter b decreases as $\exp(-2gT)$. Here we consider yet another case, namely,

$$R = igA - pA|A|^2. \tag{6.48}$$

It qualitatively describes linear amplification of a soliton and its nonlinear damping. The first equation (6.46) now has the form

$$\frac{db}{dT} = 2gb - \frac{4}{3}pb^3, \tag{6.49}$$

with the solution

$$b = \frac{b_0 e^{2gt}}{\sqrt{\frac{2pb_0^2}{3g}\left(e^{4gt} - 1\right) + 1}}, \tag{6.50}$$

where $b_0 = b(T = 0)$. If the initial amplitude is small, at the initial stage the soliton amplitude increases exponentially, whereas at large T it is saturated to the limiting value, $b_s = \sqrt{3g/2p}$, which is independent of the initial amplitude. For more examples of such "autosolitons", see Chapter 8.

The NSE has numerous variants and generalizations, such as a relevant Gross–Pitaevskii equation describing processes in the Bose–Einstein condensates [2, 27].

6.4. Rotational KdV Equation

6.4.1. *Terminal damping of solitons*

Rather distinctive is the behavior of nonlinear waves in the following non-integrable, fourth-order system [24]:

$$u_t + 3uu_x + (1/4)u_{xxx} = (1/2)\mu^2 w, \quad w_x = u. \tag{6.51}$$

The left-hand side of this equation corresponds to the KdV equation (with slightly differently defined coefficients). The small right-hand side is responsible for, e.g., the small effect of Earth's rotation on internal and surface gravity waves in the ocean, where w is the component of fluid velocity transversal to the wave propagation direction; see the review in

[13]. A similar equation was derived for the magneto-acoustic solitons in rotating plasma [23]. We call it the RKdV equation.[a] As shown in [19] and [6], despite the fact that Eq. (6.51) is non-dissipative, stationary solitary waves are, strictly speaking, impossible in this system at $\mu \neq 0$ because the term with w produces radiation resulting in a slow damping of the initial solitary wave. Here we consider how this damping develops. Note that the term with w is given the order of μ^2. Indeed, here the radiation is the direct cause of soliton amplitude variation rather than a second-order "side" effect of the first-order factors such as viscous dissipation.

A specific feature of localized (as well as spatially periodic) solutions of (6.51) is that they satisfy the constraint $\int_{-\infty}^{\infty} u dx = 0$ [24]. The corresponding regularization of the solutions of (6.51) allowing them to be matched with the basic hydrodynamic equations which do not have this limitation was discussed in [10]. We suppose that at t = 0 (and, hence, at all t > 0) the above constraint is satisfied, so that the Cauchy problem is posed correctly.

Further in this section we mainly follow the paper [11]. We seek the solution in the form

$$u = U(\zeta, T) + \mu^2 u^{(1)}(\zeta, T) + \dots,$$
$$w = W(\zeta, T) + \mu^2 u^{(1)}(\zeta, T) + \dots \tag{6.52}$$
$$\zeta = x - \int_0^t V dt, \ V = A.$$

Here the main term U is the KdV soliton, and W satisfies the second Eq. (6.51):

$$U = A \operatorname{sech}^2\left(\sqrt{A}\zeta\right),$$
$$W = \int_{\infty}^{\zeta} U d\zeta = -\sqrt{A}\left[1 - \tanh\left(\sqrt{A}\zeta\right)\right]. \tag{6.53}$$

[a] In literature it is often referred to as the "Ostrovsky equation" (e.g., [10, 12, 20, 30]).

It was assumed that W vanishes at $\zeta \to \infty$ together with U.

In the first approximation we have for the perturbation $u^{(1)}$:

$$-Vu_\zeta^{(1)} + 3(Uu^{(1)})_\zeta + \frac{1}{4}u_{\zeta\zeta\zeta}^{(1)} = -U_T + \frac{1}{2}W, \quad w_\zeta^{(1)} = u^{(1)}. \tag{6.54}$$

Using the first equation (6.9), we obtain the orthogonality condition

$$\int_{-\infty}^{\infty} U(U_T - \frac{1}{2}V)d\zeta = 0, \tag{6.55}$$

or

$$\frac{1}{2}\frac{d}{dT}\int_{-\infty}^{\infty} U^2 d\zeta + \frac{1}{4}\left(\int_{-\infty}^{\infty} U d\zeta\right)^2 = 0. \tag{6.56}$$

Substituting the soliton from (6.53), we have

$$\frac{2}{3}\frac{dA^{3/2}}{dT} + A = 0. \tag{6.57}$$

The solution of this equation is

$$A(T) = \frac{1}{4}(T_0 - T)^2. \tag{6.58}$$

The parameter $T_0 = 2\sqrt{A_0}$, where $A_0 = A(T = 0)$, is the extinction time. The corresponding soliton trajectory is

$$X_s(T) = \int_0^T A(T')dT' = \frac{1}{12}\left[T_0^3 - (T_0 - T)^3\right]. \tag{6.59}$$

Hence, the soliton disappears in a finite period of time, $T = T_0$, at the distance $X_{s0} = T_0^3/12 = 4A_0/3$. This effect may be called "terminal damping". One can say that at $T > T_0$ all soliton energy is consumed by radiation.

It is important to note that the aforementioned "antisoliton theorem" stating the non-existence of a soliton is not applicable in the case of $\mu^2 < 0$ which can also be physically sound. Various solitons and multi-solitons possible in this case were constructed numerically in [5, 13].

6.4.2. *Radiation*

Consider now the small, slowly varying field behind the soliton at times when the soliton still exists. First, one should determine a quasi-stationary, near-field part, which generally follows the soliton but includes the perturbations $u^{(1)} = u^{(1)}(\zeta, T)$ and $w^{(1)} = w^{(1)}(\zeta, T)$, which in this area can be comparable with or even exceed the soliton field in the basic approximation. Note first that $U(\zeta \to \pm\infty) \to 0$, $W(\zeta \to \infty) \to 0$, but $W(\zeta \to -\infty) \to -2\sqrt{A} \neq 0$. Then from (6.54) it follows that $u^{(1)}$ and $w^{(1)}$ tend to zero at $\zeta \to \infty$ but not behind the soliton, at $\zeta \to -\infty$. Thus, at a slow scale, the perturbation has a "jump" in the area of the soliton center, as was described in Chapter 5 (see Eq. (5.12)).

In view of the above, Eq. (6.51) for slowly varying perturbations $u^{(1)}$ and $w^{(1)}$ at large negative ζ is reduced to the following:

$$Vu_Z^{(1)} + \frac{1}{2}W + \frac{1}{2}w^{(1)} = 0, \; w_Z^{(1)} = u^{(1)}. \tag{6.60}$$

Here $Z = \mu\zeta = X - \int_0^T VdT'$. Differentiating (6.60) with respect to Z and substituting $V = A$, we obtain

$$Au_{ZZ}^{(1)} + \frac{1}{2}u^{(1)} = -\frac{1}{2}U \approx -\sqrt{A}\delta(Z). \tag{6.61}$$

Here δ is the Dirac delta function. At this scale the soliton is infinitely narrow; thus, the last equality here is the result of matching the main soliton asymptotics at $Z \to -\infty$ (where U is exponentially small) and the slowly varying perturbation at $Z \to 0$. The solution of (6.61) at $Z \leq 0$ is

$$u^{(1)} = \sqrt{2}\sin\frac{Z}{\sqrt{2A}},$$
$$w^{(1)} = \int_\infty^Z u^{(1)}dZ' = 2\sqrt{A}\left(1 - \cos\frac{Z}{\sqrt{2A}}\right). \tag{6.62}$$

The full values of u and w in this region are defined by the sums of (6.62) and the limiting values of (6.53) at $\zeta \to -\infty$:

$$u = \sqrt{2}\sin\kappa\zeta, \quad w = -2\sqrt{A}\cos\kappa\zeta, \quad \kappa = \mu / \sqrt{2A}. \tag{6.63}$$

This quasi-stationary, near-field solution covers the distances from the soliton defined as $\Delta/\mu^2 >> |\zeta| >> \Delta$, where Δ is the characteristic soliton width.

Consider now the radiated "tail" at still larger distances where the field is free in the sense that it is a non-steady function of slow variables X and T. In this area the direct effect of the soliton is negligible, and from Eq. (6.51), after neglecting the nonlinear term and the third derivative of u, we have

$$\frac{\partial u^{(1)}}{\partial t} = \frac{\mu^2}{2}w, \text{ or } \frac{\partial^2 u^{(1)}}{\partial T \partial X} = \frac{1}{2}u^{(1)}. \tag{6.64}$$

Equation (6.64) corresponds to Eq. (6.21) for the KdV case but here the perturbation is of the second order. It is a linear equation with dispersion, so that any initially localized perturbation will spread out, forming a frequency-modulated wave as discussed in Chapter 3. For a harmonic wave in the form of $\exp[i(KX - \Omega T)]$, the dispersion equation is

$$\Omega = \frac{1}{2K}, \; c_p = \frac{\Omega}{K} = \frac{1}{2K^2}, \; c_g = \frac{d\Omega}{dK} = -\frac{1}{2K^2}, \tag{6.65}$$

where c_p is the wave phase velocity and c_g is the group velocity; note that the latter is negative, so that in this frame of reference the linear wave packets move backwards. Hence, in the physical variables they propagate more slowly than the long linear waves; this confirms the above assumption that no radiation exists ahead of the soliton.

Since the "tail" is quasi-harmonic, one can seek a solution in the form $u^{(1)} = B(X,T)\exp[i\theta(T,X)]$ and use the results of Chapters 3 and 4. It follows from them that each wave group with a given $K = \theta_X$ propagates with the constant group velocity along the characteristics, or the "space–time rays" on the (T, X) plane:

$$X - c_g(K)T = F(K). \tag{6.66}$$

This is a simple envelope wave similar to that given by Eq. (3.51). The function $F(K)$ can be found from the boundary condition at the soliton trajectory, $dx/dt = dX/dT = V(T) = A(T)$. Near the soliton the radiation should be matched with the near-field perturbation (6.63) which propagates with the soliton velocity. Thus, the phase velocity of the radiated tail near the soliton is equal to the latter's velocity, $c_p = A$, $c_g = -A$. Consider a fixed ray which at a moment T_i leaves the soliton located at $X_s(T_i)$ and having the amplitude $A(T_i)$ (which is equal to its local velocity V). Then in (6.66), $F = X_s(T_i) + A(T_i)T_i$, and the characteristic (6.66) has the form

$$X = X_s(T_i) - A(T_i)(T - T_i). \tag{6.67}$$

The straight-line rays beginning at different moments $T_i \geq 0$ diverge as shown in Fig. 6.5. At a given moment T the radiated waves cover an interval $-X_f < X < X_s$; since $c_g(X_s) = -V$, we have $X_f = X_s$ which is defined by (6.59). As a result, at $T > 0$ the interval covered by the wave broadens symmetrically from the initial point $X = 0$. At the terminal moment $T = T_0$, this interval is $2X_f = 4A_0^{3/2}/3$. At $T > T_0$ the soliton disappears and the radiation propagates as a free field.

Since the soliton amplitude decreases, the wave number of the radiated wave, $K = 1/\sqrt{2A}$, increases, so that at a given moment the wavelength is maximal at the far end of the tail, near $X = -X_f$.

Asymptotic behavior of the radiated wave at $T >> T_0$ is such that $c_g = |X|/T$ (see Chapter 4) and, hence, $K = \sqrt{T/2|X|}$.

Note that in the final expressions one can substitute T and X with the initial variables t and x; in this case the small parameter μ will explicitly appear in the expressions for the radiation amplitude and wave number.

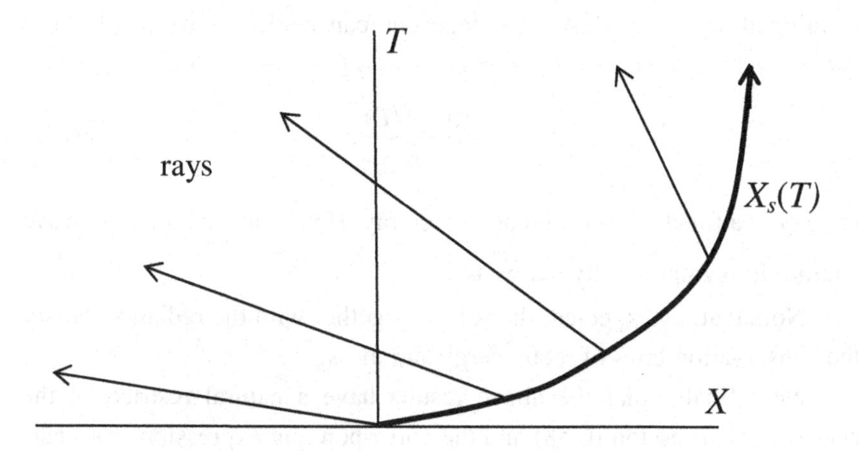

Figure 6.5. Schematic plot of the soliton trajectory $X_s(T)$ and the radiated space-time rays.

The wave amplitude can be found from the energy balance equation:

$$E_T + (c_g E)_X = 0. \tag{6.68}$$

In the linear wave $E = f(K)a^2$, where a is the wave amplitude and f is some function. Since K, as well as $c_g(K)$ and any smooth $f(K)$, propagate as a simple wave (6.66), we have

$$f_T + c_g f_X = 0, \tag{6.69}$$

and from (6.68) it follows that

$$a_T + c_g a_X + (c_g)_X a = 0. \tag{6.70}$$

According to (3.55), it has a solution:

$$a = \frac{s(F)}{\sqrt{1 + T dc_g / dF}}, \tag{6.71}$$

where F corresponds to (6.66) and s is an arbitrary function which should again be determined from the boundary condition at the soliton trajectory. We omit obtaining a cumbersome general relation $c_g(F)$ and consider only the asymptotic tail behavior at large T and X, where, as mentioned, $c_g = -X/T$. At this stage one can neglect unity in (6.71) to have

$$a = \frac{\psi(|X|/T)}{\sqrt{T}}. \tag{6.72}$$

Here ψ is a function constant at a given ray. Hence, at each ray, the wave amplitude asymptotically varies as $T^{-1/2}$.

Note that, as expected, the soliton together with the radiation satisfy the conservation laws of total energy and mass.

Note finally that the above results have a natural restriction: the zero-order expression (6.58) and the corresponding expressions for near-field perturbations are valid when the soliton amplitude A remains much larger that the tail wave amplitude. It can be shown that this condition is met when $T_0 - T >> \sqrt{\mu}$. Under the same condition, the tail wavelength at the soliton trajectory remains much longer than the characteristic length of the soliton, as it was supposed above.

These results were confirmed in [11] by numerical modeling of Eq. (6.51). Figure 6.6 illustrates the entire process.

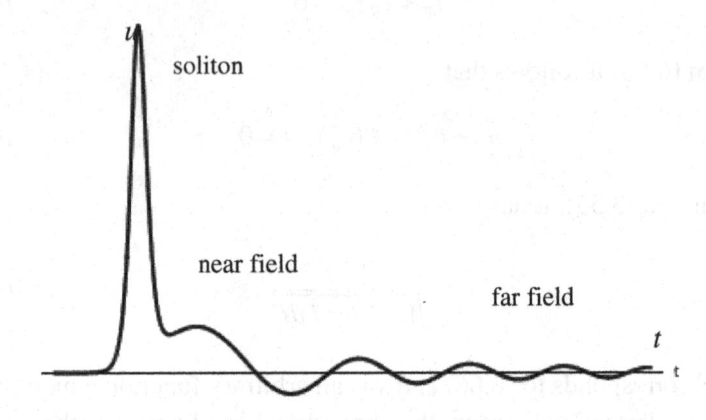

Figure 6.6. Schematic of a radiating solitary wave: time dependence of u at a fixed point.

6.5. Refraction of Solitons

6.5.1. *Geometrical theory of solitons*

So far in this chapter we have considered one-dimensional processes. Now we shall briefly describe what can be called the "geometrical theory" of solitons [26]. Here we consider a two-dimensional configuration and describe a "soliton front" represented as the curved line of the soliton's top (or bottom, depending on its polarity). By doing so we apply the approach first suggested by Whitham [26] for shock wave fronts, to the case of another localized entity, a soliton. In this section we use this "geometrical" approach together with the soliton energy conservation. As mentioned above, this approach corresponds to the main approximation of the asymptotic method.

First, introduce an orthogonal set of curvilinear coordinates, α and β, defined by the successive positions of the soliton front (α = const) and the lines perpendicular to the front, which can be called the rays (β = const). This configuration is shown in Fig. 6.7 for two successive time moments. One can interpret α as time, so that the displacement of the front from t to $t + dt$ is equal to $V(\alpha,\beta)d\alpha$, where V is soliton velocity, which varies with time as well as along the front.

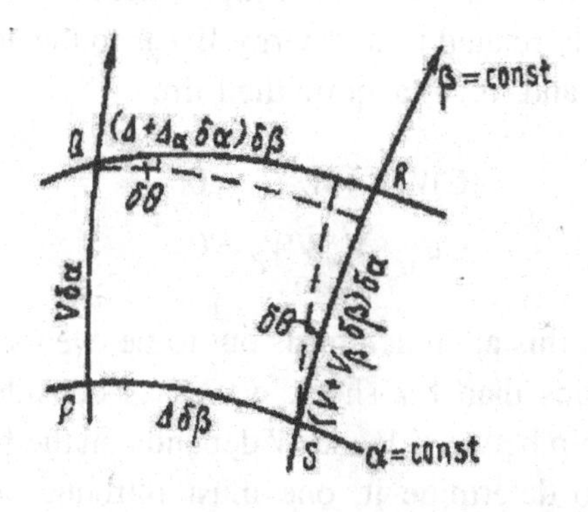

Figure 6.7. Configuration of soliton fronts and rays.

Considering the curvilinear quadrilateral $PQRS$ shown in Fig. 6.7, one can readily determine the change in the ray direction angle θ in the course of propagation. Namely, variation of the ray slope between points P and S is $\delta\theta = (QR - PS)/PQ = (1/V)(\partial\Delta/\partial\alpha)\delta\beta$, or

$$V\frac{\partial\theta}{\partial\beta} = \frac{\partial\Delta}{\partial\alpha}. \tag{6.73}$$

Here Δ is the width of the ray tube, i.e., the transverse distance between two close rays. Since the slope of a front is $\theta + \pi/2$, we similarly obtain

$$\Delta\frac{\partial\theta}{\partial\alpha} = -\frac{\partial V}{\partial\beta} \tag{6.74}$$

To close this system, one must specify the relationship between V and Δ. This relationship follows from the conservation of soliton energy:

$$W\Delta = W_0\Delta_0 = C, \tag{6.75}$$

where W is the soliton energy per unit length of its front, and subscript 0 refers to the initial front position, $\alpha = \alpha_0$. Since the scale is arbitrary, we choose it to make all the ray tubes carry the same energy. Then the constant C in (6.75) is independent of β.

For simplicity we consider an isotropic medium where the soliton velocity V is uniquely related to its energy W (or to the soliton amplitude A). Then Eqs (6.73) and (6.74) acquire the form

$$\begin{aligned} CW_\alpha + VW^2\theta_\beta &= 0, \\ C\theta_\alpha + V_W W W_\beta &= 0. \end{aligned} \tag{6.76}$$

By virtue of (6.75), this approach turns out to be even simpler and more consistent for solitons than for shock waves as considered in [31] for which the relationship between V and W depends on the flow behind the shock wave, and to determine it, one must introduce some additional simplifying assumptions.

6.5.2. *Transverse stability of a soliton*

The system (6.76) has two families of characteristics on the (α, β) plane:

$$\frac{d\beta}{d\alpha} = \pm \frac{\sqrt{WVV_W}}{\Delta}. \tag{6.77}$$

The type of system (6.76) is thus determined by the sign of the product VdV/dW; in what follows we suppose that $V > 0$. If $dV/dW > 0$, the characteristics (6.77) are real, and the system is hyperbolic. In this case Eq. (6.77) defines the velocity at which small perturbations propagate along the soliton front.

In the opposite case, when $dV/dW < 0$, the system (6.76) is elliptic, which means the instability of a plane wave: small perturbations grow exponentially. The physical interpretation is that in a stable soliton, a primarily focusing section of the front tends to straighten (V increases at the axis where the energy density increases); in the opposite, unstable case the curvature of the focusing section will grow cumulatively. The latter case is similar to the self-focusing of nonlinear wave beams in nonlinear optics [1]. Note that, as will be demonstrated below, the inequality $dV/dW > 0$ is not necessarily equivalent to the condition $dV/dA > 0$ (which is commonly used in, e.g., nonlinear optics), because the soliton energy depends not only on the amplitude A but also on its length.

As an example, consider the following modification of the KdV equation:

$$u_t + u_x + u^p u_x + u_{xxx} = 0, \quad p > 0. \tag{6.78}$$

Here the KdV equation is written in a "physical" form, with the term u_x added. In this case the velocity of a long linear wave is $c_0 = 1$, so that the soliton velocity is $V = 1 + V^{(1)}$, where $V^{(1)}$ is a small addition due to nonlinearity and dispersion which was considered as the soliton velocity in (6.1). In this section we shall use this full velocity for the description of front motion, in order to avoid problems with the cylindrical wave where the ray tube width and, hence, wave amplitude, already vary in the linear limit.

Solitary solutions of (6.78) having a plane front can easily be found from the corresponding ODE, in the same way as was done for the KdV in the beginning of this chapter:

$$u(\zeta = x - Vt) = \frac{A}{\left[ch(\zeta / \Delta)\right]^{2/p}}, \ \Delta = \frac{p\sqrt{V^{(1)}}}{2} = \sqrt{\frac{p^2 A^p}{2(1+p)(2+p)}}. \qquad (6.79)$$

At $p = 1$ this, of course, coincides with the KdV soliton (6.2) in the corresponding frame of reference.

In the lowest approximation, the soliton energy W is proportional to $A^2\Delta \sim A^{2-p/2}$, where A is the soliton amplitude and Δ is its characteristic length. Since in (6.77) $V^{(1)} \sim A^p \sim W^{2p/(4-p)}$, then

$$dV / dW = Q(p)(4-p)^{-1} W^{(3p-4)/(4-p)}, \qquad (6.80)$$

where Q is positive if $p > 0$.

Thus, at $p < 4$ a soliton is stable with respect to the considered class of perturbations; evidently, this includes the classical KdV with $p = 1$; this case was considered in application to the known Kadomtsev–Petviashvili (KP) equation which can be used for the description of a KdV soliton with a slightly curved front [15]. However, at $p > 4$ the soliton is unstable: with the increase of amplitude and velocity, it narrows so fast that the soliton energy decreases. In this case $V_W < 0$ and, according to (6.77), small initial "ripples" of the soliton front increase exponentially. In the intermediate case of $p = 4$, soliton energy is independent of amplitude in the first approximation, and a more detailed analysis is necessary to solve the stability problem.

According to (6.80), the characteristics (6.77) for perturbations propagating along the plane front have the form

$$\beta - \beta_0 = \pm\sqrt{\frac{Q}{C(4-p)}} W^{4/(4-p)}(\alpha - \alpha_0) = \pm B(W)(\alpha - \alpha_0). \qquad (6.81)$$

The signs $\pm$ correspond to propagation along the front in opposite directions. In a stable case, $p < 4$, the velocity of propagation V increases with soliton energy. In an unstable case, $p > 4$, the perturbation growth rate (or the imaginary part of the velocity) decreases with W but increases with soliton amplitude A.

In general, Eqs (6.76) describe the evolution of any bended soliton front, as long as it is smooth compared with the soliton's length across the front. In the hyperbolic case (stable soliton), these equations belong to the so-called hydrodynamic type and they can be studied by using the characteristics (6.77) on which the following two functions, Riemann invariants, are conserved:

$$J_{\pm} = \theta \mp \int \sqrt{\frac{V_W}{W}} dW = \theta \pm h(p) W^{p/(4-p)}. \tag{6.82}$$

Here $h(p)$ is defined by C, i.e., by the initial soliton amplitude or energy.

For each sign in (6.82) this process is similar to a simple (Riemann) wave:

$$W = \Phi[\beta \mp B(W)\alpha], \tag{6.83}$$

where the function Φ should be defined by the initial condition. Any point of the profile of this wave propagates with a constant velocity along the soliton front, so that the wave can eventually be overturned. An analogous effect for a shock front was referred to in [31] as "shock-shock". In our case it may be a "shock-soliton" process when a sharp change in W and θ can travel along the soliton front.

6.5.3. *Circular fronts. Nonlinear self-refraction of solitons*

Besides the planar solution, Eqs (6.76) were used in [26] to describe the evolution of a circular soliton front:

$$\begin{gathered} \theta = \beta,\ 0 \le \beta \le 2\pi, \\ \Delta(\alpha) = \frac{C}{W} = \int_0^{\alpha} (1 + V^{(1)}) d\alpha' \approx c\alpha. \end{gathered} \tag{6.84}$$

Here, we neglected the velocity perturbation V under the integrals (but not in dV/dW in (6.77)), and assumed that $\alpha = \alpha_0 \pm (t - t_0)$, where the upper and lower signs refer to the diverging and converging waves, respectively. Note that in this case the ray tube width Δ is proportional to the distance from the symmetry center.

In the case of (6.84), the characteristics (6.77) have the form

$$\beta - \beta_0 = \pm \frac{\sqrt{Q(4-p)}}{p}\left(\alpha_0^{-p/(4-p)} - \alpha^{-p/(4-p)}\right). \tag{6.85}$$

Here Q is the same as in (6.80).

For a divergent wave, unlike the case of a plane front, β remains finite at $\alpha \to \infty$. In other words, a local perturbation spreads only within a finite angle.

Consider now a somewhat more complicated example. At the initial moment, $\alpha_0 = R_0/V_0$, the soliton has a circular front with curvature radius R_0, in some angular segment, and plane elsewhere; i.e.,

$$\begin{aligned} &\theta(\alpha = \alpha_0) = \beta \ if \ -\theta_0 < \beta < \theta_0, \\ &\theta(\alpha = \alpha_0) = -\theta_0 \ if \ \beta < \theta_0, \\ &\theta(\alpha = \alpha_0) = \theta_0 \ if \ \beta > \theta_0. \end{aligned} \tag{6.86}$$

It is also supposed that the initial distribution of the energy W_0 along the entire front is uniform, $W_0(\theta) = \text{const}$. Let us first consider a convergent wave (Fig. 6.8a). In a linear case, the circular part of the wave would be focused into a point $\alpha = 0$. The characteristics of the nonlinear system (6.76) are shown in Fig. 6.8b. As it is common for hyperbolic systems, different regions of motion can be distinguished: I are regions of "rest", where the front remains plane; II are regions of simple waves, where one of the Riemann invariants (6.82) is constant, so that the solution has the form (6.83); and III is the region where the front is circular, so that the solution retains the form (6.86).

Figure 6.8b illustrates the entire focusing process. As α decreases from α_0, plane and circular regions are separated by expanding simple wave regions, where the front is less curved than in the central region. Subsequently, for a certain $\alpha = \alpha_p > 0$, the circular region vanishes completely; instead, a straight segment appears in the central region of the front; in other words, defocusing takes place. Later, for some $\alpha = \alpha_*$, the characteristics begin to intersect in region II, which corresponds to a front breaking.

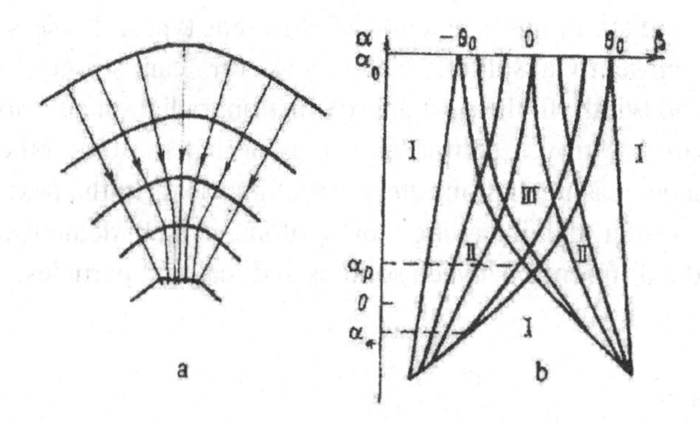

Figure 6.8. Focusing of a cylindrical soliton. a) rays and fronts, b) characteristics in the (α, β) plane.

The breaking stage needs a more detailed description that takes into account higher-order effects, such as diffraction. A dispersion relation extending (6.77) to the accounting of diffraction was obtained by Shrira [29]. Using the results of Kadomtsev and Petviashvili [15], a Burgers-type equation for relatively small but nonlinear perturbations propagating along the soliton front was obtained in the form:

$$\varphi_\alpha \pm V(\varphi_0)\left(\varphi_\beta + \frac{4}{p\varphi_0}(\varphi - \varphi_0)\varphi_\beta\right) - \sqrt{\frac{2}{3}}\varphi_{\beta\beta} = 0, \qquad (6.87)$$

where $\varphi = \int dV/(u(V)\Delta(V)$, $u = \sqrt{-V/(\Delta\Delta_V)}$. The last term in the left-hand side of (6.87) is responsible for energy losses due to radiation from the curved soliton. These finite-amplitude perturbations can propagate similarly to the Taylor shock wave in a compressive medium which exists due to the medium viscosity or thermal conductivity. A more detailed analysis of waves propagating along the soliton front can be found in the aforementioned papers.

6.4. Concluding Remarks

This chapter demonstrated examples of perturbation theory application to soliton evolution under the action of perturbations such as small dissipation, radiation, and refraction of different types. These solutions are locally close to a soliton which, however, can strongly vary in amplitude and width. In the next approximation, radiation appears which can eventually destroy a soliton and transform it to some other wave form or to a decreasing and infinitely stretching field. In the next chapter we shall consider the interaction of solitons, which demonstrates an analogy and a difference between solitons and material particles.

References

1. Boyd, R. (2008). *Nonlinear Optics*, 3rd Ed. Academic Press, London.
2. Carretero-Gonzalez, R., Frantzeskakis, D. J., and Kevrekidis, P. G. (2008). Nonlinear waves in Bose–Einstein condensates: Physical relevance and mathematical techniques, *Nonlinearity*, v. 21, pp. R139–R202.
3. Chow, V. T. (1959). *Open-Channel Hydraulics*. McGraw-Hill Inc., New York.
4. Courant, R. and Friedrichs, K. O. (1977). *Supersonic Flow and Shock Waves.* Springer–Verlag.
5. Fraunie, P. and Stepanyants, Y. (2002). Decay of cylindrical and spherical solitons in rotating media, *Physics Letters A,* v. 293, pp. 166–172.
6. Galkin, V. M. and Stepanyants, Y. A. (1991). On the existence of stationary solitary waves in a rotating fluid, *Journal of Applied Mathematics and Mechanics,* v. 55, pp. 939–943.
7. Gorshkov, K. A. (2007). *Perturbation Theory in Soliton Dynamics.* Doctor of Science Thesis, Institute of Applied Physics, Nizhny Novgorod, [in Russian].
8. Gorshkov, K. A. and Papko, V. V. (1977). Non-adiabatic stage of soliton damping and intermediate asymptotics, *Radiophysics and Quantum Electronics*, v. 20, pp. 360–365.
9. Grimshaw R. (1979). Slowly varying solitary waves. I. Korteweg–de Vries equation, and II. Nonlinear Schrödinger equation, *Proceedings of the Royal Society*, v. 368A, pp. 359–375 and pp. 377–388.
10. Grimshaw, R. H. J. (1999). Adjustment processes and radiating solitary waves in a regularised Ostrovsky equation, *European Journal Mechanic, B-Fluids,* v. 18, pp. 535–543.
11. Grimshaw, R. H, J., He, J. M., and Ostrovsky, L.A. (1998). Terminal damping of a solitary wave due to radiation in rotational systems, *Studies in Applied Mathematics*, v.101, pp. 197–210.

12. Grimshaw, R., and Helfrich, K. R. (2008). Long-time solutions of the Ostrovsky equation, *Studies in Applied Mathematics*, v. 121, pp. 71–88.
13. Grimshaw, R. H. J., Ostrovsky, L. A., Shrira, V. I., and Stepanyants, Y. A. (1998). Long nonlinear surface and internal gravity waves in a rotating ocean, *Surveys in Geophysics*, v. 19, pp. 289–338.
14. Hasegawa, A. and Kodama, Y. (1995). *Solitons in Optical Communications*. Clarendon Press, Oxford.
15. Kadomtsev, B. B. and Petviashvili, V. I. (1970). On the stability of solitary waves in weakly dispersive media, *Soviet Physics Doklady*. v. 15, pp. 539–541.
16. Kaup, D. J. and Newell A. C. (1978). Soliton as particle, oscillator and in slowly changing media: a singular perturbation theory, *Proceedings of the Royal Society, London A*, v. 301 (1701), pp. 413–446.
17. Kivshar, Y. S. and Agrawal, G. A. (2003). *Optical Solitons*. Elsevier: Academic Press, San Diego.
18. Landau, L. D. and Lifshits, E. M. (1987). *Fluid Dynamics*. Pergamon, New York.
19. Leonov, A. I. (1981). The effect of Earth rotation on the propagation of weak nonlinear surface and internal long oceanic waves, *Annals of the New York Academy of Science*, v. 373, pp. 150–159.
20. Li, Q. and Farmer, D. M. (2011). The generation and evolution of nonlinear internal waves in the deep basin of the South China Sea, *Journal of Physical Oceanography*, v. 41, pp. 1345–1363.
21. Malomed, B. A.(2006). *Soliton management in periodic systems*. Springer, New York.
22. Miles, J. W. (1983). Wave evolution over a gradual slope with turbulent friction, *Journal of Fluid Mechanics*, v. 133, pp. 207–216.
23. Obregon, M. A. and Stepanyants, Yu. A. (1998). Oblique magneto-acoustic solitons in rotating plasma. *Physics Letters A*, v. 249, pp. 315–323.
24. Ostrovsky, L. A. (1978). Nonlinear internal waves in a rotating ocean, *Okeanology*, v. 18, pp. 119–125.
25. Ostrovsky, L. A. and Potapov, A. I. (1999), *Modulated waves. Theory and Applications*. Johns Hopkins University Press, Baltimore.
26. Ostrovsky, L. A. and Shrira, V. I. (1976). Instability and self-refraction of solitons, *Soviet Physics JETP*, v. 44, pp. 738–743.
27. Pitaevskii, L. P., and Stringari, S. (2003). *Bose–Einstein Condensation*. Clarendon Press, Oxford.
28. Scott, A. C., Chu, F. Y., and McLaughlin, D.W. (1973). The soliton: A new concept in applied science, *Proceedings of IEEE*, v. 6, 1443–1483.
29. Shrira, V. I. (1980). Nonlinear refraction of solitons, *Soviet Physics JETP*, v. 52, 44–54.
30. Stepanyants Y.A. (2006). On stationary solutions of the reduced Ostrovsky equation: Periodic waves, compactons and compound solitons. *Chaos, Solitons and Fractals*, v. 28, pp. 193–204.
31. Whitham, G. B. (1974). *Linear and Nonlinear Waves*. Wiley, New York.

Chapter 7

Interaction and Ensembles of Solitons and Kinks

Sometimes my pencil is more clever than I am.

Leonhard Euler

From the previous chapter we could see that a soliton can preserve its individuality and remain close to the family of solitary solutions even if its amplitude and velocity vary significantly under the action of a perturbation. In this chapter we consider other examples of the behavior of solitons as stable entities, interacting with each other in a way that is much similar to collisions of material particles. As already mentioned, the term "soliton" stems from the computational result by Zabusky and Kruskal [38] showing that two solitary waves, after their nonlinear interaction, can eventually restore their parameters as if they were elastically colliding particles, so that only their total delays (phases) are changed. Since then, an exact description of such type of soliton interactions has been obtained for dozens of nonlinear equations, with the use of such methods such as the inverse scattering method, Bäcklund transform, and the Hirota method (e.g., [29]). Along with the exact methods, the perturbation methods can be used as was discussed in Chapter 5. This is especially important for a broad class of non-integrable systems which are not treatable by exact methods. Moreover, even for the equations close to integrable, the exact solutions are often cumbersome and non-transparent, impeding general qualitative understanding of the process.

In this chapter we apply the perturbation theory described above to the analysis of soliton interaction in a rather general class of systems allowing the Lagrangian formulation; this form of the equations is not critical for our purposes but it often allows a more elegant representation of the results. Namely, in the first non-trivial order, an ensemble of weakly interacting solitons can be described by a Lagrangian similar to that of an ensemble of classical particles with a paired interaction potential. Thus, in the lowest approximation a "classical mechanics" of solitons can be formulated, and their interactions can be classified as attractions and repulsions, which allows the qualitative description of the main types of motion for interacting solitons and, in particular, formulating the conditions for their bound states. At the same time, this approach allows one to describe a possible radiation from interacting solitons. Comparison with the available exact solutions confirms the validity of the approximate solutions.

Besides the localized solitons and their ensembles ("chains" or "lattices"), we shall also describe interaction and ensembles of the transitional, kink-type waves.

This chapter is admittedly longer and somewhat more cumbersome than others, but the main ideas and the approach used in it remain as simple as elsewhere in this book.

7.1. General Scheme

First, we shall adapt the perturbation method outlined in Chapter 5 to the problem of soliton interaction [9]. We consider a "rarefied" ensemble of solitons separated at distances larger than the characteristic scale of each soliton. In other words, each soliton is affected by weak remote parts ("tails") of others (Fig. 7.1). To preserve the "rarefaction" for a sufficiently long time, soliton velocities $\mathbf{V}_i$ should not strongly differ. As a result, distances between solitons can vary strongly only at the space–time scales much larger than the characteristic lengths of individual solitons. This allows one to again introduce slow variables $T = \mu t$ and $\mathbf{X} = \mu \mathbf{r}$ with the small parameter

$$\mu \sim \left|\mathbf{V}_{\max} - \mathbf{V}_{\min}\right| / |\mathbf{V}|, \quad \mathbf{V} = \left(\sum_{i=1}^{N} \mathbf{V}_i\right) / N, \tag{7.1}$$

where N is the number of interacting solitons, $\mathbf{V}_{max}$ and $\mathbf{V}_{min}$ are maximal and minimal soliton velocities in the group, and $\mathbf{V}$ is the average velocity.

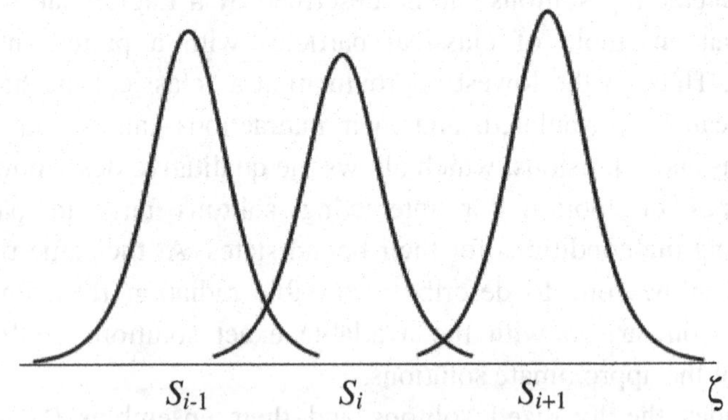

Figure 7.1. Scheme of solitons interacting by their "tails".

In the vicinity of a selected i-th soliton, the solution can be sought in the form

$$\mathbf{u}_i(t,\mathbf{r}) = \mathbf{U}(\zeta - \mathbf{S}_i, \mathbf{V}) + \sum_n \mu^n \mathbf{u}_i^{(n)}\left(\zeta - \mathbf{S}_i, T, \mathbf{X}\right), \tag{7.2}$$

where the main term $\mathbf{U}$ is a non-perturbed solitary solution depending on $\zeta = \mathbf{r} - \mathbf{V}t$ and tending to constants at infinity: $\mathbf{U}(\zeta \to \pm\infty, \mathbf{V}) = \mathbf{U}_{\pm}$. Thus, in the main order all solitons are identical, except for the coordinates of their centers, $\mathbf{S}_i = \mathbf{S}_i(T)$. The differences emerge in the dependencies $\mathbf{S}_i(T)$, as well as in the next-order perturbations.

Before dwelling on formalities, we note that neighboring solitons exchange their energy because each soliton propagates on the variable "tails" of other solitons. In most integrable systems (such as the KdV) the rear soliton slows down while the frontal one accelerates, because the energy is transferred from the rear to the frontal soliton. As a result, the solitons are "repulsed" and tend to diverge. In the opposite case, when the rear soliton accelerates, they are "attracted". We shall demonstrate this in more detail below.

To describe this process, we use the matched asymptotic expansions. Namely, the series (7.2) will be used for constructing the solution in the vicinity of the soliton ("internal" region) and then matching it with the slowly varying solution in the "external" regions between solitons.

As in Chapter 5, perturbations in each order of μ satisfy the linear equations:

$$G\mathbf{u}_i^{(n)} = \mathbf{H}_i^{(n)}, \tag{7.3}$$

where G is a partial-derivative operator obtained by linearization of the basic equations, as in Eq. (5.4). As before, the functions $\mathbf{H}^{(n)}$ depend on the perturbations found in previous approximations. It is to be remembered that, notwithstanding the specific form of G, it depends only on the zero-approximation function $\mathbf{U}$.

In the context of the matching procedure, let us analyze perturbations in the external area where the main solution is close to its constant asymptotics, $\mathbf{U}(|\zeta - \mathbf{S}_i|) = \mathbf{U}_{ex} = \text{const}$ which is, as mentioned, the same for all solitons; in particular, it can be $\mathbf{U}_{ex} = 0$. The solution in this area can be represented by series analogous to (7.2) where the main term is $\mathbf{U}_{ex}$. To match the expansions in the areas between the solitons we expand each zero-order solution at large $|\zeta|$ in an asymptotic series around $\mathbf{U}_{ex}$:

$$\mathbf{U}_i(\zeta, \mathbf{V}) = \mathbf{U}_{iex} + \sum_n \mu^n \mathbf{f}^{(n)}(\zeta, \mathbf{V}), \tag{7.4}$$

where the perturbations $\mathbf{f}^{(n)}$ can be found from the linearized equations similar to (7.3). However, since $\mathbf{U}_{ex} = \text{const}$, here the operator G is autonomous (has constant coefficients). Also, in this area $\mathbf{H}^{(1)} = 0$ because it contains only the derivatives of $\mathbf{U}_{ex}$. Hence, the main asymptotics are to be found from the homogeneous equation

$$G\mathbf{u}^{(1)} = 0. \tag{7.5}$$

The series (7.4) converges within some interval $|\zeta| < r_0 < \infty$. Here r_0 defines the minimal distance from the center of the i-th soliton at which the matching can be made. Actually r_0 is of the order of minimal distance between the neighboring solitons for which this approximate approach is adequate. In the one-dimensional case, when (7.5) is a linear ODE with constant coefficients (in the two-dimensional and three-dimensional cases the coefficients, in general, depend on geometrical parameters, such as radius r), in many cases the soliton asymptotics are exponents: $\mathbf{u}^{(1)}(\zeta,\mathbf{V}) \sim \exp(-\lambda(\mathbf{V})|\zeta|)$. It includes such important equations as KdV, Klein–Gordon, and the nonlinear Schrödinger equations considered above. In practice, the convergence interval r_0 can even be of the order of the soliton length, so that the corresponding approximate solutions can remain close to exact ones even when the solitons overlap.

The expansions (7.4) include both the asymptotics of the main solution $\mathbf{U}$ and perturbations caused by the "tails" of other solitons. To enable a correct matching, we assume that the main asymptotics of each soliton have the order of μ in the external domain where the matching takes place, and the order of μ^2 near the centers of neighboring solitons. Certainly, the solutions also depend on the basic soliton parameters such as the velocity $\mathbf{V}$.

This general outline will now be detailed for the equations in the Lagrangian form.

7.2. Lagrangian Description

The behavior of solitons as particles is especially well demonstrable in the Lagrangian representation of the field that was briefly discussed in Chapter 5. Following [9] and, in additional details [7], we consider the Lagrangian equations (5.13) with $\mathbf{R} = 0$ when the perturbations are exerted by one soliton on another:

$$\frac{\partial}{\partial t}\frac{\partial L}{\partial \mathbf{u}_t} + \frac{\partial}{\partial \mathbf{r}}\frac{\partial L}{\partial \mathbf{u}_\mathbf{r}} - \frac{\partial L}{\partial \mathbf{u}} = 0. \tag{7.6}$$

To make somewhat cumbersome formulae more compact, in this section we consider the one-dimensional problem when both $u(x, t)$ and $\zeta = x - Vt$ are scalar variables. Generalization of the perturbation scheme to the two-dimensional and three-dimensional problems needs a simple change of scalar variables to the vector ones. Such an example will be considered in the end of this chapter.

The basic solution satisfies the ODE (5.14):

$$\frac{d}{d\zeta}\left(\frac{\partial L}{\partial U_x}-V\frac{\partial L}{\partial U_t}\right)-\frac{\partial L}{\partial U}=0. \tag{7.7}$$

According to the above, the solution in the vicinity of a chosen soliton is represented in the form (7.2), and perturbations satisfy the linear equations (7.3), namely,

$$Gu^{(n)}=\frac{d}{d\zeta}\left[\begin{array}{l}\left(V^2\frac{\partial^2 L}{\partial U_t^2}+\frac{\partial^2 L}{\partial U_x^2}-2V\frac{\partial^2 L}{\partial U_t\partial U_x}\right)u_\zeta^{(n)}\\ +\left(-V\frac{\partial^2 L}{\partial U\partial U_t}+\frac{\partial^2 L}{\partial U\partial U_x}\right)u^{(n)}\end{array}\right] \tag{7.8}$$
$$-\left[\left(-V\frac{\partial^2 L}{\partial U\partial U_t}+\frac{\partial^2 L}{\partial U\partial U_x}\right)u_\zeta^{(n)}+\frac{\partial^2 L}{\partial U^2}u^{(n)}\right]=H^{(n)}.$$

As before, the functions $H_i^{(n)}$ depend on the perturbations known from previous approximations. In particular, $H_i^{(1)}$ contains only U and its derivatives with respect to the slow time T, so that $\partial/\partial T=-(\partial/\partial\zeta)(dS_i/dT)$:

$$H_i^{(1)}=\frac{dS_i}{dT}\left\{\frac{\partial}{\partial\zeta}\frac{\partial L}{\partial U_t}-\left[\frac{\partial}{\partial\zeta}\left(V\frac{\partial^2 L}{\partial U_t^2}-\frac{\partial^2 L}{\partial U_x\partial U_t}\right)+\frac{\partial^2 L}{\partial U\partial U_t}\right]U_\zeta\right\}. \tag{7.9}$$

A particular solution of the linear system (7.3) with the "forcing" (7.9) is

$$u_i^{(1)}=\frac{dS_i}{dT}\cdot\frac{\partial U_i}{\partial V}. \tag{7.10}$$

To verify that this is a solution, one should differentiate (7.7) with respect to V, which yields

$$-\frac{\partial}{\partial\zeta}\frac{\partial L}{\partial U_t}+\left[\frac{\partial}{\partial\zeta}\left(V\frac{\partial^2 L}{\partial U_t^2}-\frac{\partial^2 L}{\partial U_x\partial U_t}\right)+\frac{\partial^2 L}{\partial U\partial U_t}\right]U_\zeta=G\left(\frac{\partial U}{\partial V}\right).$$

After multiplying by dS_i/dt, the left-hand side of this equation coincides with the expression (7.9) for $H^{(1)}$. Hence, (7.10) satisfies Eq. (7.8) at $n = 1$.

As a result, the full multi-soliton solution up to the first approximation has the form

$$u(x,t) = U + \mu u^{(1)} = \sum_i \left[U(\zeta - S_i, V) + \mu \frac{dS_i}{dT} \cdot \frac{\partial U}{\partial V}(\zeta - S_i, V) \right]. \tag{7.11}$$

In this order of μ the solution remains a superposition of solitons with unperturbed shapes but with shifted coordinates $S_i(T)$. Thereby the orthogonality conditions (5.15) of Chapter 5 are still identically met, and no independent equations defining variation of soliton coordinates $S_i(T)$ can be obtained in the first approximation. To describe the soliton dynamics, one has to consider the second approximation. As mentioned, the asymptotics of second-order perturbation in the region adjacent to an i-th soliton consists of the perturbation of its own field, $u_i^{(2)}$, and the first-order asymptotics of other solitons which are supposed to be of the second order in the considered region:

$$u^{(2)}(x,t) = u_i^{(2)}(\zeta - S_i, T, X) + \sum_{j \neq i} u^{(1)}(\zeta - S_j, V), \tag{7.12}$$

where $u^{(1)}$ is the first-order term in the series (7.4) which satisfies (7.5). Of main interest here is the function $u_i^{(2)}$ responsible for the orthogonality condition. Equation (7.3) for this part has the form

$$Gu_i^{(2)} = H_i^{(2)}(U_i, u_i^{(1)}) - \sum_{j \neq i} Gu_i^{(1)}(\zeta - S_j, V). \tag{7.13}$$

Correspondingly, the function $H^{(2)}$ can be presented in the form $H_i^{(2)} = \left(H_i^{(2)}\right)_1 + \left(H_i^{(2)}\right)_2$, where

$$\left(H_i^{(2)}\right)_1 = \frac{dS_i}{dT}\left[\frac{\partial H_i^{(1)}}{\partial V} - \frac{dS_i}{dT}\frac{\partial L}{\partial V} \cdot \frac{\partial U_i}{\partial V}\right], \tag{7.14}$$

$$\left(H_i^{(2)}\right)_2 = \frac{d^2 S_i}{dT^2}\left[\frac{\partial^2 L}{\partial U_t^2}(U_i + VU_i) - \frac{\partial^2 L}{\partial U_t \partial U_x} U_{i\xi V} - \frac{\partial^2 L}{\partial U_t \partial U} U_{iV}\right.$$
$$\left. + \frac{\partial}{\partial \zeta}\left(\frac{\partial^2 L}{\partial U_t^2} VU_{iV} - \frac{\partial^2 L}{\partial U_x \partial U_t} U_{iV}\right) + \frac{\partial^2 L}{\partial U \partial U_t} U_{iV}\right]. \tag{7.15}$$

The expression (7.14) contains a second-order addition to soliton velocity, $(dS_i/dT)^2$, whereas (7.15) is proportional to the soliton acceleration, d^2S_i/dT^2. It is easy to see that (7.14) is a localized function of $\zeta - S_i$. In fact, it represents a second-order addition to the localized solution (7.11) which remains quasi-stationary and localized. To show this one should, as above, multiply (7.8) and (7.9) by dS_i/dT at $n = 1$ and then differentiate with respect to V.

However, the function $\left(H_i^{(2)}\right)_2$ contains an exponentially growing perturbation $u_i^{(2)}$, and to keep it finite, the following orthogonality condition should be met:

$$\int_{-\infty}^{\infty} d\zeta\, U_{i\zeta}\left[\left(H_i^{(2)}\right)_2 - \sum_{j\neq i} G u_j^{(1)}\left(\zeta - S_j\right)\right] = 0. \tag{7.16}$$

The equations for soliton coordinates follow from here:

$$m\frac{d^2 S_i}{dT^2} = \sum_{j\neq i} F\left(S_j - S_i\right). \tag{7.17}$$

Here m is defined by integrating the first term in (7.16) with $\left(H_i^{(2)}\right)_2$ given by (7.15):

$$m = \left(\frac{d^2 S}{dT^2}\right)^{-1} \int_{-\infty}^{\infty} U_\zeta H_1^{(2)} d\zeta = \frac{\partial P}{\partial V}, \quad P = \int_{-\infty}^{\infty} U_\zeta \frac{\partial L}{\partial U_t} d\zeta. \tag{7.18}$$

Here P is the full wave momentum of the soliton field, so that m has the meaning of the physical mass.

The function F is obtained by integrating the second term in (7.16):

$$F(S_i - S_j) = \int_{-\infty}^{\infty} U_\zeta G u_j^{(1)} d\zeta = -\frac{\partial}{\partial S_i} W(S_i - S_j), \tag{7.19}$$

where

$$W(S_i - S_j) = \int_{-\infty}^{\infty} d\zeta \left[\frac{d}{d\zeta}\left(-V\frac{\partial L^{(N)}}{\partial U_t} + \frac{\partial L^{(N)}}{\partial U_x} \right) - \frac{\partial L^{(N)}}{\partial U} \right] \cdot U(\zeta + S_i - S_j). \tag{7.20}$$

Here $L^{(N)}$ is the part of L responsible for nonlinearity, $L^{(N)} = L - L_q$, where L_q is a possible quadratic form of U and its derivatives. Indeed, for $L = L_q$ the variational equations (7.6) are linear, and L_q does not contribute to interaction. After that, the second-order perturbation $u^{(2)}$ remains bounded everywhere. It can contain a part localized near the soliton, as well as a non-localized part which can be responsible for radiation. Note that due to the conservative character of Lagrangian equations (7.6), the function F is antisymmetric, so that $F(S_i - S_j) = -F(S_j - S_i)$. As a result, Eqs (7.17) have the form of Newtonian equations for solitons as classical particles having each the mass m (let us remember that all solitons have close parameters) and interacting via the potential W. Correspondingly, (7.17) can be obtained from the effective Lagrangian with the pair potential W:

$$L_{eff} = \frac{m}{2}\sum_i \left(\frac{dS_i}{dT} \right)^2 - \sum_{j \neq i} W(S_i - S_j). \tag{7.21}$$

7.3. Types of Soliton Interactions: Repulsion, Attraction, and Bound States

The behavior of soliton "tails" can be defined from the linearized equations describing stationary waves, which, as mentioned, in most cases have an exponential or, in general, a multi-exponential form[a]:

$$u^{(1)}(\zeta) = \sum_m C_m \operatorname{Re} e^{\lambda_m \zeta}, \tag{7.22}$$

[a] Non-exponential but sufficiently strongly decreasing asymptotics, such as that in the "algebraic" Benjamin–Ono soliton, can be considered in a similar way.

where C_m are constants and λ_m are the characteristic exponents to be found from the linear equations with constant coefficients obtained by linearization of the basic equation in the areas far from the soliton center; evidently, only the exponents decreasing at $\zeta \to \pm\infty$ should be taken into account. The expression (7.22) implies the exponential character of W and F above; indeed, integrating in (7.20) can only change the coefficients C_m. Especially simple is the rather typical case when there is only one exponent in (7.22), so that for a pair of interacting solitons we have

$$m\frac{d^2S}{dT^2} = -2W'(S) = \alpha \operatorname{Re}\left(e^{-\lambda S}\right), \quad S = S_2 - S_1, \ \alpha = \text{const}. \quad (7.23)$$

Three qualitatively different cases can be distinguished here [9]:

1. $\operatorname{Im}(\lambda) = 0$ (i.e., the asymptotic potential energy profile $W(\zeta)$ is monotonous) and $\alpha m > 0$. This case corresponds to the repulsion of solitons. Integrating (7.22) we have

$$S = \pm\left[\frac{S_{\min}}{2} + \frac{1}{\lambda}\ln\cosh\left(\frac{\lambda S_{T\infty}}{2}T\right)\right], \quad (7.24)$$

where

$$S_{\min} = \lambda^{-1} \cdot \ln\left[\frac{2\alpha}{\lambda m S_{T\infty}^2}\right] \quad (7.25)$$

is the minimal achievable distance between solitons if they approach each other from a large distance at the initial velocity $S_{T\infty}$.

As follows from (7.24), after interaction the solitons exchange their velocities: $\dot{S}_{1,2}(t \to +\infty) = \dot{S}_{2,1}(t \to -\infty)$, so that the only non-trivial result of collision is a shift of their coordinates by

$$\Delta x = \pm\lambda^{-1} \cdot \ln\left(\frac{\alpha}{2\lambda m S_{T\infty}^2}\right). \quad (7.26)$$

The signs $\pm$ in (7.24) and (7.26) refer to the faster and the slower soliton respectively. Note that if $m > 0$, the energy is pumped from the rear to the frontal soliton. This effect has a "parametric" interpretation: the field of the rear soliton is increasing the velocity of the frontal one, and vice versa.

2. $\mathrm{Im}(\lambda) = 0$ but $am < 0$. In this case solitons are attracted to each other. Even if the solitons initially diverge, then, from some maximal distance S_{max}, they begin closing in up to the distance of the order of their width, λ^{-1}, when their fields begin to overlap, and our approximation ceases to work. If, however, after close interaction the solitons separate again, the process repeats itself. As a result, an excited bound state, or a breather, occurs, with an oscillating motion of solitons. If the time of strong overlapping is much smaller than the period τ of these oscillations, the latter can be found from (7.23) as

$$\tau = 8\left(\frac{1}{2\alpha\lambda}\frac{\partial P}{\partial V}\right)^{1/2}\exp\left(\frac{\lambda S_{max}}{2}\right)\arccos\left[\exp\left(-\frac{\lambda S_{max}}{2}\right)\right]. \tag{7.27}$$

3. If λ is complex, $\mathrm{Im}\lambda \neq 0$, the soliton "tail" oscillates, so that the interaction potential contains a sequence of minimums and maximums with a decreasing magnitude and a spatial period of the order of $(\mathrm{Im}\lambda)^{-1}$. This case allows both infinite motions and bound states of solitons.

Figure 7.2 illustrates the three cases outlined above.

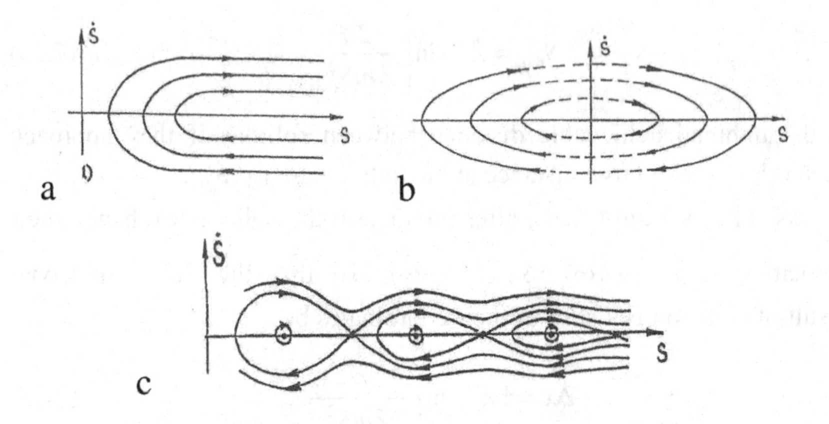

Figure 7.2. Qualitative phase portraits of Eq. (7.23) for two interacting solitons with monotonous (a – repulsing, b – attracting) and oscillating (c) asymptotics.

If the soliton asymptotics consist of several exponents, the solution of (7.23) can still always be represented in quadratures. In this case the main role in the process is played by the longest exponent (with the minimal $\mathrm{Re}\lambda_m$).

7.4. A Generalized KdV Equation

As an example, consider an equation

$$u_t + u^p u_x + u_x^{(2q+1)} = 0. \tag{7.28}$$

This is a further generalization of KdV from (6.78) in which $q = 1$.

Unlike (6.78), in general this equation is not integrable and solitons can not always be represented in an explicit analytical form. However, a sufficiently complete analysis is still possible. A solitary solution of this equation at any p and q (if it exists) has a self-similar form:

$$U = V^{1/p} f\left(V^{1/2q}\zeta\right), \quad \zeta = x - Vt. \tag{7.29}$$

Here the function f is localized.

This equation can be represented in Lagrangian form. In particular, for $q = 1$ and $q = 2$ the corresponding Lagrangian is

$$L = \frac{1}{2}v_x v_t + \frac{1}{(p+1)(p+2)}(v_x)^{p+2} + \left\{ \begin{array}{ll} v_x w_x + \frac{1}{2}w^2 & \text{if } q = 1, \\ v_x w_x + wu + \frac{1}{2}u_x^2 & \text{if } q = 2 \end{array} \right\}, \tag{7.30}$$

where the notations are: $u = v_x$, $w = v_{xx} = u_x$.

Consider the interaction of two solitons. From (7.18) and (7.20) it follows that

$$m = \frac{\partial P}{\partial V} = \frac{\partial}{\partial V}\int_{-\infty}^{+\infty} d\zeta \, \frac{\partial L}{\partial v_t^{(0)}} \cdot v_x^{(0)} = \frac{1}{2}\frac{\partial}{\partial V}\int_{-\infty}^{+\infty} d\xi \, U^2(\zeta),$$

$$W(S) = \int_{-\infty}^{+\infty} d\zeta \, \frac{\partial L^{(N)}}{\partial v_x^{(0)}} v^{(0)}(\zeta - S) = \frac{1}{p+1}\int_{-\infty}^{+\infty} d\zeta \, (U(\zeta))^{p+1} u^{(1)}(\zeta - S). \tag{7.31}$$

In the case of $q = 1$ and any $p > 0$ which was considered in Chapter 6, there exists an explicit solution (6.79) for a soliton, so that the integrals in (7.31) give

$$\frac{\partial P}{\partial V} = (4-p)p^2\left[(1+p)(2+p)\right]^{2/p} \cdot 2^{(2/p)-2} \cdot \Gamma^2\left(\frac{2}{p}\right) \cdot \Gamma^{-1}\left(\frac{4}{p}\right) \cdot V^{(2/p)-(3/2)},$$

$$W'(S) = 2^{(2/p)+1}\left[(1+p)(2+p)\right]^{2/p} \cdot V^{(2/p)+1} \cdot \exp\left(-\sqrt{V}S\right). \tag{7.32}$$

Here Γ is gamma-function. It is seen that at $p < 4$ we have $\partial P / \partial V > 0$, so that the solitons are mutually repulsed. In this case the expression for the minimal distance between the solitons corresponding to (7.26) is

$$\Delta x = \frac{1}{\sqrt{V}} \ln \left[\frac{2p^2 \cdot \Gamma\left(\frac{4}{p}\right) \cdot V^2}{\left(1 - \frac{p}{4}\right) \cdot \Gamma^2\left(\frac{2}{p}\right) S_{T\infty}^2} \right]. \tag{7.33}$$

In particular, for $p = 1$ (KdV case), the distance S between the solitons satisfies the equation

$$\frac{d^2 S}{dT^2} = \pm 32 V^{3/2} \exp(-V^{1/2} S). \tag{7.34}$$

This equation has an analytical solution. According to (7.33), the minimal distance between the solitons in this case is

$$\Delta x = \frac{2}{\sqrt{V}} \ln\left(4V / S_{T\infty}\right). \tag{7.35}$$

It is interesting that the same condition, $p < 4$, defines the two-dimensional transverse stability of a soliton front as described in the previous chapter.

At $p > 4$ solitons are attracted and, as described above, an oscillating bound state is possible. In this case the soliton is transversely unstable. As mentioned, physically it is related to the fact that at $p > 4$, the soliton length decreases so rapidly with the increase of its amplitude that its energy and momentum also decrease.

Note also that at even p, solitons can have both positive and negative polarity. When solitons of opposite signs interact, their interaction potential is attracting, and bound states of solitons are possible already at $p = 2$ as in the well-known MKdV equation where an oscillating pair of solitons, a breather, is possible. The same result follows from the exact solution of MKdV [25].

Now consider the case of $p = 1$, $q = 2$. This is a non-integrable case, and there is no analytic expression for the function f in (7.29). A single solitary solution was numerically studied by Kawahara [23], who showed that "slow" solitons ($V < 0$) have oscillating asymptotics which can be found by linearization of the stationary equation; namely,

$$u^{(1)}(\zeta) \sim \mathrm{Re}\exp\left(\frac{1+i}{2}\sqrt[4]{V}\zeta\right). \tag{7.36}$$

Taking into account the general structure of the solution shown in (7.29), it is possible to represent Eq. (7.23) as

$$\frac{d^2S}{dT^2} = \alpha\exp\left(-\frac{\sqrt[4]{V}}{2}S\right)\cdot\cos\left(\frac{\sqrt[4]{V}}{2}S\right), \tag{7.37}$$

where

$$\alpha = \frac{9\sqrt{2}}{4} V \exp\left(\frac{\sqrt[4]{V} S_0}{2} \right) \cdot \left(A^2 + B^2 \right)^{1/2} \cdot C^{-1} \cdot D,$$

$$S_0 = \arctan\left(A / B \right), \; A = \left\langle f^{\,2}(z) e^{-z} \cos z \right\rangle, \tag{7.38}$$

$$B = \left\langle f^{\,2}(z) e^{-z} \sin z \right\rangle, \; C = \left\langle f^{\,2}(z) \right\rangle, \left\langle ... \right\rangle = \int_{-\infty}^{+\infty} ...dz,$$

and D is determined from the condition $f\left(z \to \infty\right) = D \cdot e^{-z} \cos z$.

The dynamics of these solitons can be understood from Fig. 7.2c. Some trajectories continue to infinity; at the corresponding initial condition the solitons, after reaching a minimal distance, then infinitely diverge. However, there are regions with closed trajectories, corresponding to finite motions, which represent bound states with periodic, oscillating motions of the soliton pair. Also, there exist equilibrium points corresponding to a stationary localized solution with two maximums, i.e., a "two-hump" soliton. Moreover, there is an infinite (albeit countable) number of such solutions. This result obtained in the first approximation is confirmed by the analysis of the four-dimensional space of the stationary equation corresponding to (7.28) with $p = 1$ and $q = 2$ showing that there exists a homoclinic structure with an infinite, countable number of trajectories describing such multi-solitons [13]. From the latter it follows that at $p = 1$ and $q = 2$, Eq. (7.28) is non-integrable. Note also that different multi-solitons can be of interest as possible classical analogs of various elementary particles in the nonlinear field theory.

An important question is that of radiation from soliton interaction. Indeed, the interacting solitons perturb each other and, according to the results of Chapter 6, they can radiate. On the other hand, in the integrable equations, the total radiation does not exist. This question was briefly considered in [10] for the integrable KdV and the non-integrable Zakharov equations. In both cases each soliton radiates during interaction. However, in the integrable case, radiation "tails" of two solitons compensate each other everywhere except for the region between them, and in the latter area the radiation decreases when solitons diverge after interaction. In the non-integrable case, there is no such compensation, and the radiation penetrates outside the interaction interval.

7.5. Soliton Lattices and their Stochastization

Based on the above consideration, we shall now discuss the dynamics of soliton ensembles in the form of quasi-periodic sequences of weakly interacting solitons. If solitons with close parameters are well separated and interact by their asymptotics ("tails"), one can take into account only interactions of a given (i-th) soliton with its closest neighbors from both sides. In this case, the system (7.17) describes a chain (lattice) of coupled nonlinear oscillators:

$$m\frac{d^2S_i}{dT^2} = W'(S_i - S_{i-1}) - W'(S_{i+1} - S_i). \tag{7.39}$$

Consider first the perturbation of soliton positions in this chain which are much smaller than the distances between them but can be comparable to soliton lengths. Here we again suppose that the solitons have exponential asymptotics, $W'(S) = a\exp(-\lambda S)$, so that (7.39) has the form

$$m\frac{d^2S_i}{dT^2} = \alpha\left[e^{-(S_i - S_{i-1})} - e^{-(S_{i+1} - S_i)}\right]. \tag{7.40}$$

In a strictly periodic soliton lattice, the right-hand side of (7.40) is evidently zero, so that the lattice is stationary. Considering small deviations of solitons from their equilibrium positions, it is easy to see that the lattice is stable if $am > 0$ and unstable if $am < 0$. This has a simple physical explanation: according to (7.23), a lattice of repulsing solitons is stable, whereas a lattice of attracting solitons is unstable. Note that if the lattice modulation is slow, so that the displacements of neighboring solitons are close to each other, the differences can be replaced by differentials, so that (7.40) becomes a wave equation:

$$m\frac{\partial^2 S(n,T)}{\partial T^2}=\alpha\frac{\partial^2 S(n,T)}{\partial n^2}+\alpha\left(\frac{\partial S(n,T)}{\partial n}\right)^2. \quad (7.41)$$

Here n is the number of a soliton in the chain. This equation is hyperbolic at $am > 0$ and elliptic at $am < 0$. This is similar to the results of the theory for quasi-periodic waves discussed in Chapter 3. In the framework of (7.41), nonlinear wave evolution ends up with the formation of stepwise singularities (wave "breaking") in the hyperbolic case, and the formation of a cusp in the elliptic case [27]. In general, however, Eq. (7.40) is discrete and, consequently, dispersive. The dispersion prevents singularities and, in particular, allows the existence of regular stationary waves in such a soliton lattice.

In the stable case, (7.40) coincides with the known Toda lattice equation [37]. As known, it has an exact solution in the form of a periodic stationary wave:

$$S_n-S_{n-1}=\Lambda_0+\lambda^{-1}\ln\left\{1-(2K\nu)^2\left[dn^2 2K(\nu t-kn)-\frac{E}{K}\right]\right\}, \quad (7.42)$$

where dn is an elliptic function (Jacobi delta-amplitude), E and K are complete elliptic integrals with modulus k, and the velocity ν is related to k by a nonlinear dispersion relation:

$$2K\nu=\left[sn^{-2}(2Kk)-1+EK^{-1}\right]^{-1/2}. \quad (7.43)$$

Besides periodic solutions, Eq. (7.43) includes localized solitary waves,

$$S_n-S_{n-1}=\Lambda_0+\lambda^{-1}\ln\left[1-\beta^2\cosh^{-2}\lambda_1(n-\nu t)\right], \quad (7.44)$$

where $\beta=\sinh\lambda_1$ and $\nu=\lambda_1^{-1}\sinh\lambda_1$.

With respect to the basic, periodic sequence of solitons, the solutions (7.42) and (7.44) can be considered as stationary "envelope waves" propagating with the group velocity ν, and modulating both the distances between solitons and their amplitudes. They include the "soliton-soliton" structures. In this context, (7.41) describes non-dispersive envelope waves of the so-called hydrodynamic type. The modulated "soliton-soliton" chain is illustrated in Fig. 7.3.

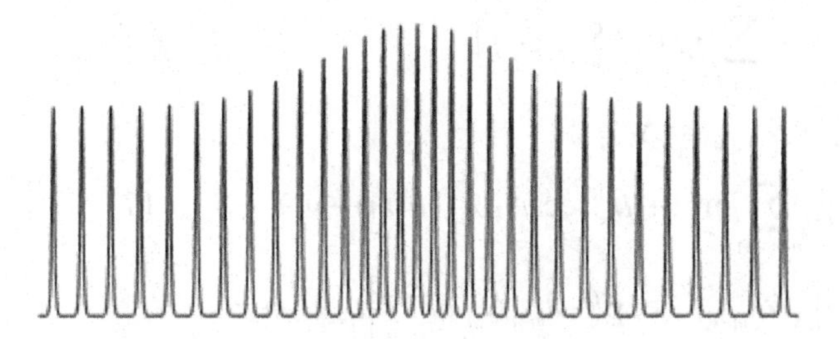

Figure 7.3. Envelope soliton modulating soliton amplitudes and their positions in a chain.

The above theory implies that, strictly speaking, the amplitudes of solitons do not vary strongly, However, as mentioned, the theory is often applicable far beyond its formal limitations.

As is known [28], the Toda system (7.40) is completely integrable, from which it follows that the envelope solitons (7.44) can interact without having their parameters changed, except for the additional phase shifts. In the particular case of KdV solitons, the exact multi-soliton solution obtained by Kuznetsov and Mikhailov [24] possess similar properties.

A remarkable feature of these solutions is that one can construct a hierarchy of envelope solitons. Indeed, the solutions (7.44) again have exponential asymptotics and, consequently, can form a lattice of mutually repulsing envelope solitons, perturbations of which are, in turn, described by Toda equations similar to (7.40), which has an "envelope-envelope" soliton similar to (7.44). Repeating this reasoning, we come to the hierarchy of different-order soliton lattices, each of which is in fact an excitation of a soliton lattice of a previous order. An analytical expression for an envelope wave system of an m-th order has the form [14, 18]

$$u(x,t) = \sum_{n_{(1)}} U\left[\zeta - S_{n(1)}(T)\right],$$
$$S_{n(m)} - S_{n-1(m)} = \Lambda_{m-1}$$
$$+\lambda_{m-1}^{-1} \sum_{n(m+1)} \ln\left\{1+\beta_m^2 \operatorname{sec} h^2\left[\lambda_m\left(n(m) - v_m T - S_{m+1,n}\left(T_{m+1}\right)\right)\right]\right\}, \quad (7.45)$$
$$T_m = \left[\left(\alpha\lambda_{m-1}\right)^{1/2} \exp\left(-\lambda_{m-1}\Lambda_{m-1}/2\right)\right] t_{m-1}, \; m = 1,...N.$$

Here $u(x, t)$ is the basic sequence of solitons, $U(\zeta = x - Vt)$, v_m and Λ_m are the velocity and period of an m-th order lattice, $n(m)$ refers to a m-th order lattice, and α is the interaction parameter in the Toda lattice (7.40). Using (7.44), the parameters β_m and λ_m can be expressed via $v_m + dS_{m,n+1}/dT$.

The solution (7.45) is a N– periodic function with not necessarily multiple periods. For integrable models they are apparently close to a particular class of exact, N-zonal solutions [6] which are, however, too cumbersome to compare in detail with the above approximate ones. On the other hand, the approximate solution (7.45) is simpler and more general in the sense that it is not related to the feature of integrability. One can say that here the Toda lattice plays a role similar to the role of the nonlinear Schrödinger equation for modulated quasi-harmonic waves.

7.6. Stable and Unstable Soliton Lattices

Another interesting topic is the interaction of solitons with non-monotonous asymptotic behavior which is rather typical of many non-integrable models. We already discussed this case for Eq. (7.28) at $p = 1$ and $q = 2$. A somewhat more general but related equation has the form

$$\frac{\partial u}{\partial t} + u\frac{\partial u}{\partial x} + \frac{\partial^3 u}{\partial x^3} + \gamma\frac{\partial^5 u}{\partial x^5} = 0, \quad (7.46)$$

A similar equation was derived in [22] for magneto-acoustic waves in plasma; now equations of the (7.46) type are known for a number of physical systems; among them are the gravity-capillary waves on thin liquid films [36] and electromagnetic waves in nonlinear lines [14, 30]. At $\gamma > 0$ solitary solutions of this equation can have oscillating tails [23].

As mentioned, at interaction this asymptotics yields the force $U'(S) \sim e^{-\lambda_1 S} \cos\lambda_2 S$. We have already seen that such solitons can form bound states. In the case of a soliton lattice, variation of its average period leads to switching between stable and unstable states, depending on the sign of $U'(S)$. In the stability zones, when the lattice oscillates moderately, the process is similar to that considered above, and an envelope soliton hierarchy can be constructed. If, however, the lattice period is such that solitons are located in the unstable zone, small initial displacements of solitons from their equilibriums initiate strong displacements tending to approach stable areas. Note that such behavior qualitatively corresponds to the Lennard-Jones potential for interacting particles. For the latter, stochastization was found in numerical experiments [5]. Thus, one can suppose that stochastization is possible for the unstable zone of Eq. (7.46) as well. This was confirmed in the experiments described briefly below.

Similar properties are demonstrated by the RKdV equation (6.54) of Chapter 6, after changing the sign in the last term in the left-hand part (otherwise there are no solitons):

$$u_t + 3uu_x + (1/4)u_{xxx} + (\gamma/2)w = 0, \quad w_x = u. \tag{7.47}$$

At $-2\sqrt{\gamma} < V < 2\sqrt{\gamma}$, linearization of this equation for soliton asymptotic areas gives, as above, oscillating decay, so that interaction between solitons may be both repulsive and attractive, depending on the distance between their maximums, i.e., on whether they are located in minimums or maximums of the potential energy of their interaction. Considering two neighboring solitons with oscillating tails, one can construct stationary solutions in the form of stable or unstable "bisolitons". Examples of such solutions numerically constructed for Eq. (7.47) in [31] (see also [18]) are shown in Fig. 7.4.

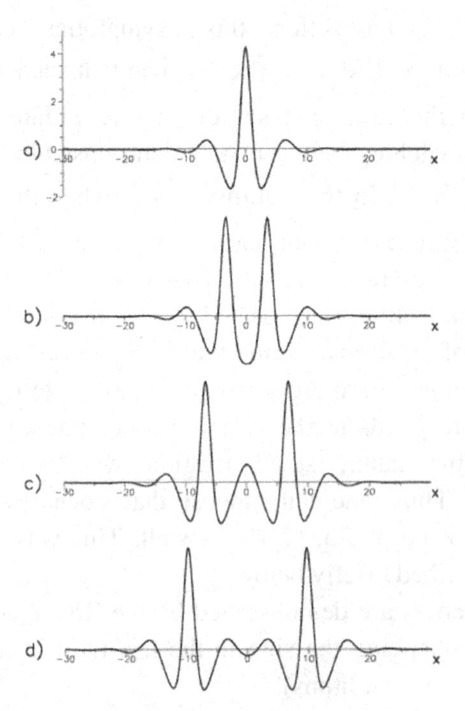

Figure 7.4. Examples of stable and unstable bisolitons for Eq. (7.47) with $\gamma = 1$.
a) a single soliton, b) and d) stable, and c) unstable bisoliton.

Stable bisolitons correspond to the cases when the maximum of one soliton is located in one of the local minimums (potential wells) of another soliton (Figs 7.4b and d), whereas unstable bisolitons correspond to the cases when the maximum of one soliton is located in one of the local maximums of another soliton (Fig. 7.4c).

The solutions in the form of bisolitons were also obtained for two-dimensional models such as the Kadomtsev–Petviashvili (KP) equation and its generalizations [1, 2]. For the KP equation a bisoliton solution and even more complex stationary multi-soliton solutions were derived analytically [34], but these constructions were found to be unstable.

7.7. Interaction of Solitons in Electromagnetic Lines

Many of the processes considered above have been studied experimentally in electromagnetic chain lines containing nonlinear capacitances and/or inductances. Such lines are, strictly speaking, described by differential-difference equations. However, for the processes with a spatial scale covering many individual elements of the chain, continual analogs of these equations can be used (compare with (7.41)). In [11] a line was described which consists of nonlinear capacitances and linear inductances coupled by additional inductive connections *M* between neighboring elements, as shown in Fig. 7.5.

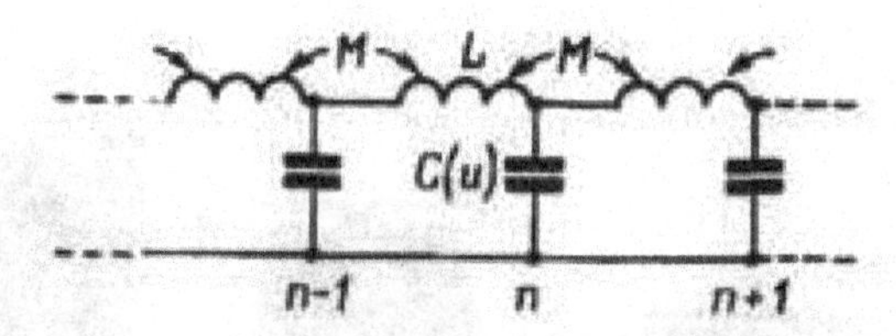

Figure 7.5. Scheme of a transmission line containing linear inductances *L* and *M*, and nonlinear capacitances *C* depending on the voltage *u*.

A continual version of the equation describing this line is an equation similar to (7.46):

$$u_t + \alpha u u_x + \beta u_{xxx} + \gamma u_{xxxxx} = 0. \tag{7.48}$$

Here the term with α appears due to a weak nonlinearity of the capacitance $C(u)$, the term with β is obtained by expansion of differences of voltages at neighboring elements, and the term with γ is due to the interconnections *M*.

Various solitons and multi-solitons were observed in this line. Some examples are shown in Fig. 7.6.

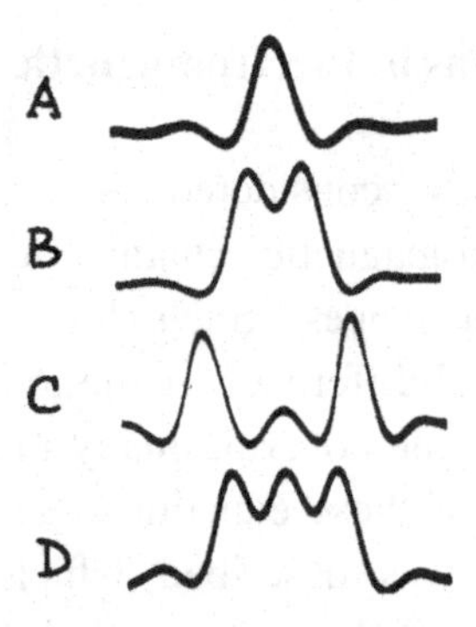

Figure 7.6. Different solitary waves in an electromagnetic line corresponding to Eq. (7.48) with the parameters $\alpha = 2$, $\beta = -0.03$, $\gamma = -0.0047$. A – one-hump soliton; B to D – stable multi-hump solitons. From [13].

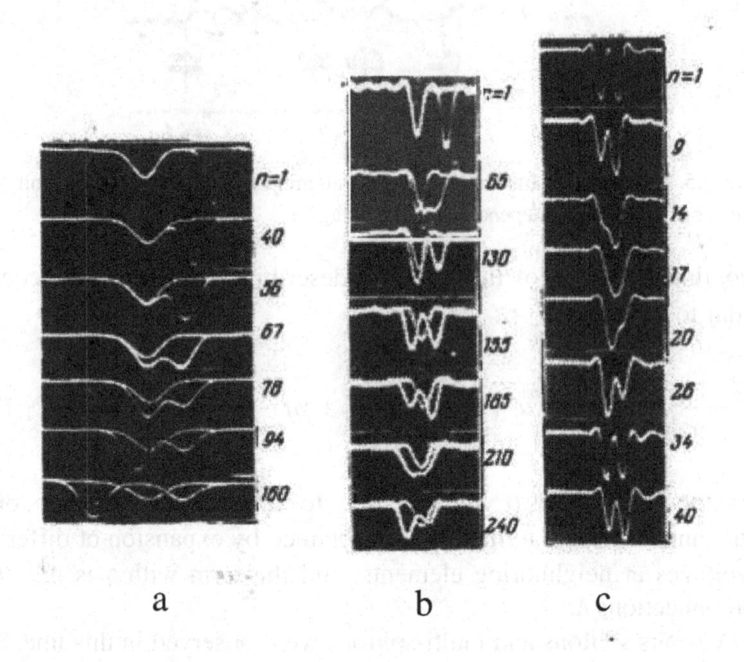

Figure 7.7. Two-soliton interaction in the nonlinear electromagnetic line, (a) without inductive coupling between elements; (b) and (c) with such coupling. In (a) and (b) the position of a single reference pulse is also shown. n is the number of the elements of the chain corresponding to the distance along the line. The abscissa is time.

Figure 7.7 illustrates the interaction of electromagnetic solitons [11]. Figure 7.7a corresponds to the KdV case when the solitons have exponential asymptotics. They are mutually repulsing and, as a result, exchange parameters after interaction. Figures 7.7b and c show the bound states of solitons with oscillating asymptotics occurring at $M \neq 0$.

Correspondingly, stable and unstable soliton lattices can exist in such systems, as illustrated in Fig. 7.8. For a stable lattice, an "envelope soliton" was observed in such a line. In unstable lattices, stochastization is possible provided the excitation energy exceeds the depth of the corresponding potential minimum in which the solitons are located. In [14], soliton lattices were excited in the line by a harmonic source which quickly formed a slightly perturbed soliton lattice. By changing the source period, it was possible to obtain both stable and unstable lattices, when the neighboring solitons are located in minimums and maximums of the interaction potential, respectively.

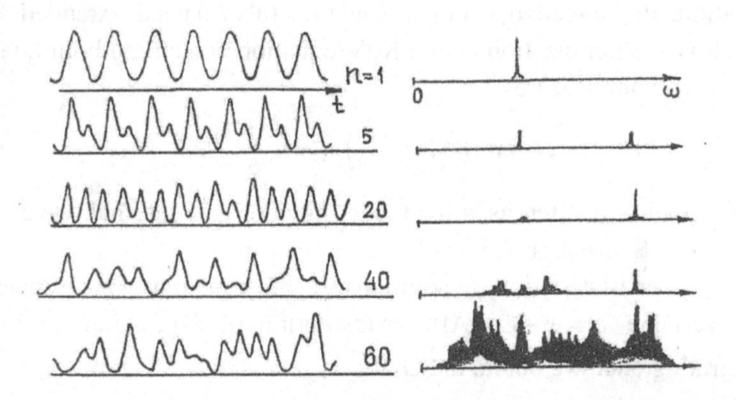

Figure 7.8. Left: development of an unstable soliton lattice along the line. The time variation of voltage was recorded at different line cells (n is cell number). Right: the corresponding wave spectra.

Figure 7.8 shows that at distances $n > 60$ the wave spectrum becomes practically continuous (in spite of the periodic excitation), which can be considered as a criterion of stochastization. This is still an ensemble of solitons, but a random one. Soliton turbulence in electromagnetic lines was studied in more detail in [12].

7.8. Interaction of Flat-Top Solitons and Kinks

7.8.1. *Compound solitons in the Gardner equation*

Consider now another class of solitary waves which are referred to as kinks. These waves are transitions between two different constant levels U_1 and U_2. It should be stressed that in this chapter we deal with non-dissipative, conservative transitions rather than with dissipative waves (shock waves) and active ones (autowaves) which will be discussed in the next chapter. There are many examples of the equations having such kinds of solutions: Sine–Gordon, MKdV, etc. In many cases kinks appear as a limiting solution for a family of stationary solutions; in the phase space they correspond to a separatrix connecting two different equilibrium points rather than returning to the same point as a "classic" soliton does. The general scheme described in Chapter 5 is applicable here with some specifics. As an example we consider an integrable equation, the so-called Gardner equation (also named extended KdV), which is a generalization of the KdV equation containing both quadratic and cubic nonlinearities:

$$u_t + 6\left(u - u^2\right)u_x + u_{xxx} = 0. \tag{7.49}$$

This equation is often used for modeling nonlinear internal waves in the ocean (see Subsection 7.7.3 below).

Consider first stationary solutions of this equation, which depend on one variable $\zeta = x - Vt$. After substitution of $U(\zeta)$ into (7.49) and integrating once we obtain the ODE:

$$U_{\zeta\zeta} - 2U^3 + 3U^2 - VU = 0. \tag{7.50}$$

Here the integration constant is let zero, so that we consider kinks tending to zero in one of their asymptotics. Figure 7.9 shows the corresponding phase trajectories of Eq. (7.50) which begin or end at the equilibrium point $U = 0$ at different velocities V.

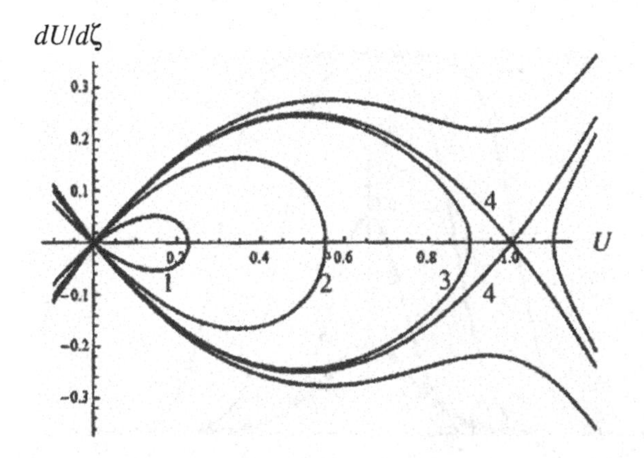

Figure 7.9. Phase trajectories of Eq. (7.50) going through the zero equilibrium point, $U=0,\ dU/d\zeta=0$, for different wave velocities. Closed trajectories (solitons) marked 1, 2, and 3 correspond to the velocities V = 0.4, 0.8, and 0.99, respectively. The separatrices 4 connecting two equilibrium points correspond to a kink and "antikink".

As seen from Fig. 7.9, the solutions of Eq. (7.50) include solitons. Analytically, such solitary solutions have the form (e.g., [21]):

$$U(x,t)=\frac{k}{2}\left[\tanh\frac{k}{2}\left(x-k^2t+\Delta\right)-\tanh\frac{k}{2}\left(x-k^2t-\Delta\right)\right], \qquad (7.51)$$

where $k=\sqrt{V}$, $\Delta=k^{-1}\text{arctanh}(k)$, and $0\le k\le 1$. In the limit of small k, this solution coincides with the KdV soliton considered above. In the other limit, when k is close to unity, the solution can be approximated as a compound of two separated kinks:

$$U(x,t)\approx\frac{k}{2}\left[\begin{array}{c}\tanh\left(\frac{k}{2}(x-k^2t)+\frac{1}{4}\ln\frac{2}{1-k}\right)\\ -\tanh\left(\frac{k}{2}(x-k^2t)-\frac{1}{4}\ln\frac{2}{1-k}\right)\end{array}\right]. \qquad (7.52)$$

Figure 7.10 shows soliton profiles for different values of the parameter $\mu=1-k$.

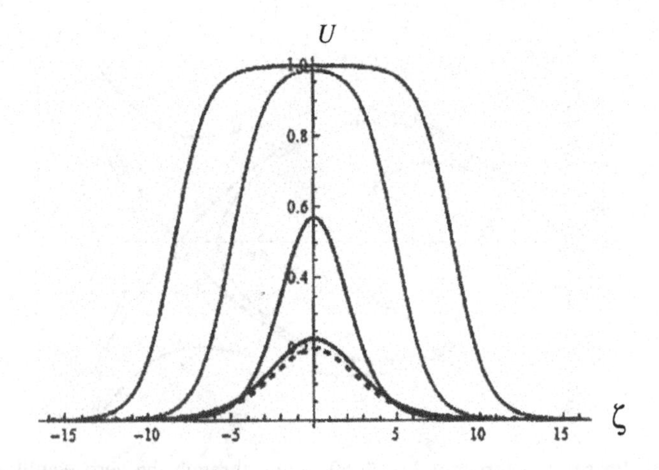

Figure 7.10. Soliton profiles at four different values of the parameter μ = 1− *k*. From the top down: μ = 10^{-6}, 10^{-4}, 0.1, and 0.4, respectively. Solid lines: exact solution (7.51). Dashed lines: approximate solution (7.52). Except for $\mu = 0.4$, exact and approximate profiles are indistinguishable.

It is seen that the difference between the exact and approximate solutions is noticeable only for the relatively small solitons, when *k* is less than, say, 0.5. For a very small μ (large *k*) we have a flat-top soliton with the edges close to kinks, each of which is an exact solution at *k* = 1:

$$U^{\pm}(x,t) = \frac{1}{2}\left[1 \pm \tanh\frac{1}{2}(x-t)\right]. \tag{7.53}$$

Here two signs correspond to two separatrices in Fig. 7.9; these two solutions can be called kink (at soliton front) and antikink (at its trailing edge).

Thus, there exists a maximal, limiting soliton amplitude, and near this maximum the soliton is close to a compound of two kinks with different polarities, separated by an interval 2*L* which is large in comparison with the characteristic kink length:

$$U(x,t)_{k\to 1} \approx U^{+} + U^{-} - 1 = \frac{1}{2}\left[\tanh\left(\frac{x-t+L}{2}\right) - \tanh\left(\frac{x-t-L}{2}\right)\right]. \tag{7.54}$$

This expression is represented in a form characteristic of the method of matched asymptotic expansions: the solution consists of two regions of rapid variation of the function U, separated by a plateau where $U \approx 1$. The form (7.54) provides an incentive to represent the main approximation for a general N-soliton solution, $U^{(N)}$, as a superposition of $2N$ kinks of alternating signs:

$$U^{(N)} = \frac{1}{2}\sum_{i=1}^{2N}(-1)^{i+1}\tanh\frac{1}{2}\left[\zeta - S_i(T,X)\right], \qquad (7.55)$$

where $\zeta = x - t$ and $S_i\,(T, X)$ is a slowly varying coordinate of the kink front. In this section the small parameter μ is chosen to be equal to $1-k$.

As a result, instead of the interaction of N solitons as unchanged entities, we consider the interaction of $2N$ kinks of alternating signs forming these solitons. Thus, an algorithm for the construction of the general solution consists of finding local solutions near each kink and their subsequent matching in the areas between the kinks [17, 19]. The solution in the vicinity of an i-th kink is represented as an asymptotic series in μ:

$$u_i(x,t) = U_i(\zeta - S_i) + \sum_{n=1}^{2N}\mu^n u_i^{(n)}(\zeta - S_i, T, X), \qquad (7.56)$$

where U_i is a kink of the corresponding sign and each perturbation $u^{(n)}$ is to be found as a function of $\zeta - S_i$ from linear ODEs that emerge, as above, after substituting (7.56) into (7.49) and expanding in powers of μ:

$$G_i u_i^{(n)} = H_i^{(n)}\left(U, \ldots\, u_i^{(n-1)}\right). \qquad (7.57)$$

Here

$$G_i = \frac{d}{d\zeta}\left[-1 + \frac{3}{2}\operatorname{sech}^2\frac{1}{2}(\zeta - S_i) + \frac{d^2}{d\zeta^2}\right], \qquad (7.58)$$

and, in particular,

$$H_i^{(1)} = \left(\frac{\partial S_i}{\partial T} + \frac{\partial S_i}{\partial X}\right) U_\zeta + 2\frac{\partial S_i}{\partial X} U_{\zeta\zeta\zeta}. \quad (7.59)$$

where U represents a kink (7.53) of the corresponding polarity.

The difference between this scheme and the one constructed above for solitons is that a soliton has at least one arbitrary parameter (such as soliton velocity V or amplitude A), whereas each kink is completely defined except for the coordinate of its front. As a result, matching of perturbations in the intervals between the kinks becomes necessary beginning from the first approximation. Similarly to what we did before, in the first approximation, the perturbations $u_i^{(1)}$ at two sides of the i-th kink are matched with the respective perturbations $u_{i-1}^{(1)}$ and $u_{i+1}^{(1)}$. Note in this connection that the decreasing asymptotics of neighboring zero-order kinks are included in the first-order perturbations:

$$U_{i-1}\left[(\zeta - S_{i-1}) \to \infty\right] \to \begin{cases} e^{(-\zeta - S_{i-1})} & \text{(odd } i\text{)} \\ 1 - e^{(-\zeta + S_{i-1})} & \text{(even } i\text{)} \end{cases},$$

and (7.60)

$$U_{i+1}\left[(\zeta - S_{i+1}) \to -\infty\right] \to \begin{cases} 1 - e^{(\zeta - S_{i+1})} & \text{(odd } i\text{)} \\ e^{(\zeta - S_{i+1})} & \text{(even } i\text{)} \end{cases}.$$

In other words, the values of $\exp(-|S_{i\pm1} - S_i|)$ are of the order of μ in the vicinity of the neighboring soliton. It is essential that the same exponents form the first-order solutions $u_i^{(1)}$ between the kinks. Indeed, according to (7.59), at constant asymptotic values of U we have $H_1^{(1)} = 0$, and in the limit of large $|\zeta|$ the operator G_i in (7.58) has constant parameters. As a result, the asymptotic solution of (7.57) has the form $u_i^{(1)} = C_{i1}e^{\zeta} + C_{i2}e^{\zeta} + C_{i3}$, where $C_{i1,2,3}$ are constants. Thus, for the asymptotic part of the solution the matching is reduced to finding these constants by using (7.60).

The condition of finiteness of the rest of the first-order perturbation can be reduced to the orthogonality condition:

$$\int_{-\infty}^{\infty}\int_{0}^{\zeta} U_{\zeta}\tilde{H}^{(1)}d\zeta' d\zeta = 0,$$
$$\text{where } \tilde{H}^{(1)} = H^{(1)} - (-1)^i G\left(e^{\zeta - S_{i+1}} - e^{-(\zeta - S_{i-1})}\right). \tag{7.61}$$

In view of (7.59), this condition can be reduced to a system of equations for the kinks' coordinates:

$$\frac{dS_i}{dT} = \frac{\partial S_i}{\partial T} + \frac{\partial S_i}{\partial X} = -2D_i + 4(-1)^i\left(e^{-(S_{i+1}-S_i)} - e^{-(S_i - S_{i-1})}\right), \tag{7.62}$$

where D_i are integration constants. Besides, the parts of the functions $u_i^{(1)}$ which do not vanish at $\zeta - S_i \to \pm\infty$, should be matched, which results in additional equations:

$$\left[1 + (-1)^{i+1}\right]\frac{d}{dT}(S_{i+1} - S_i) + 2(D_{i+1} - D_i) = 0. \tag{7.63}$$

This equation allows one to exclude the parameters D_2, D_3,…, one after another, from Eqs (7.62), and to obtain the equations for the phases of kinks:

$$\frac{dS_i}{dT} = -4\left[e^{-(S_{i+1}-S_i)} + e^{-(S_i - S_{i-1})}\right] + 2D_1. \tag{7.64}$$

Differentiating (7.64) with respect to T at a constant D_1 and excluding the terms with the first derivatives dS_i/dT, we obtain

$$\frac{d^2S_i}{dT^2} = 16\left[e^{-(S_{i+2}-S_i)} + e^{-(S_i - S_{i-2})}\right]. \tag{7.65}$$

These equations differ from the Toda equations (7.40) for solitons in that they form two independent Toda lattices for the even- and odd-numbered kinks (i.e., for soliton fronts and trailing edges). These two lattices are connected by Eqs (7.64) which represent the known Bäcklund transform for the systems (7.65). In general, these two systems can differ. Here we consider a degenerated case when $D_1 = 0$. In this case the solutions of each system can be written independently, whereas the Bäcklund transform (7.64) relates their parameters.

The global first-order perturbation $u^{(1)}$ can be represented as a sum of local kink solutions $u_i^{(1)}$ minus the sum of their common asymptotics between the kinks, $u_{i+}^{(1)} = u_i^{(1)}(\zeta - S_i)_{|\zeta|\to\infty}$. After using the system (7.64) at $D_1 = 0$ this perturbation can be written in an explicit form:

$$u^{(1)} = \sum_i^{2N}\left(u_i^{(1)} - u_{i+}^{(1)}\right) = \frac{1}{4}\sum_i (-1)^{i+1}\left[\begin{array}{l}\dfrac{dS_i}{dT}\tanh\left(\dfrac{\zeta - S_i}{2}\right) \\ +\left(\dfrac{\partial S_i}{\partial T} + 3\dfrac{\partial S_i}{\partial X}\right)\left(\dfrac{\zeta - S_i}{2}\right)\cosh^{-2}\left(\dfrac{\zeta - S_i}{2}\right)\end{array}\right]. \tag{7.66}$$

This is not yet a uniformly suitable solution. Indeed, although the last term in square parentheses exponentially decreases at large $|\zeta|$, its ratio to the zero-order solution U secularly increases in proportion to $|\zeta - S_i|$. To eliminate this growth, the following condition should be met:

$$\frac{\partial S_i}{\partial T} + 3\frac{\partial S_i}{\partial X} = 0, \quad i = 1, 2, \ldots 2N. \tag{7.67}$$

Equation (7.67) shows an interesting result: if the kink phases are slowly perturbed, this perturbation has the form $S_i(X,T) = S_i(X - 3T)$. In other words, the phase variation (an analogue of the "envelope wave" for quasi-harmonic modulated waves) propagates with the "group velocity" which, in the accepted variables, is three times larger than the velocity of a soliton with the critical amplitude ("phase velocity").

Now an N-soliton solution wich includes the first-order perturbations (i. e., takes interactions into account) can be represented as

$$U_{N_s} + u_{N_s}^{(1)} = \frac{1}{2}\sum_{i=1}^{2N}(-1)^{i+1}\left(1 - \frac{\partial S_i}{\partial X}\right)\tanh\frac{1}{2}\left[\zeta - S_i(X,T)\right]. \tag{7.68}$$

In this sum the slowly varying coefficients $(1 - \partial S_i / \partial X)/2$ are x – derivatives of the full phases of the kinks, in the same way as the factor $k/2$ is the x – derivative of the argument of hyperbolic tangent in the primary stationary soliton (7.51). This means that the expression (7.68) preserves the form of a N-soliton solution as a superposition of quasi-stationary solitons, with the corresponding shift in space. As in most of integrable systems, the interaction of Gardner solitons qualitatively develops as the collision of repulsing particles: after approaching a minimal distance, they diverge. The specifics here are due to the presence of the internal structure of Gardner solitons: when the solitons are close to each other, frontal and rear kinks can behave differently: interaction between the fronts begins earlier than that of rear kinks (see the example below).

7.8.2. *Two-soliton interaction*

Figure 7.11 illustrates the interaction of two solitons close to the tabletop ones considered each as a compound of two kinks. In this case Eq. (7.65) for the phases of the four kinks has the form

$$\frac{\partial^2 S_{1,3}}{\partial T^2} = \mp 16 e^{-(S_3 - S_1)}, \quad \frac{\partial^2 S_{2,4}}{\partial T^2} = \mp 16 e^{-(S_4 - S_2)}, \tag{7.69}$$

where the signs + and – refer to the frontal (even i) and rear (odd i) kinks, respectively.

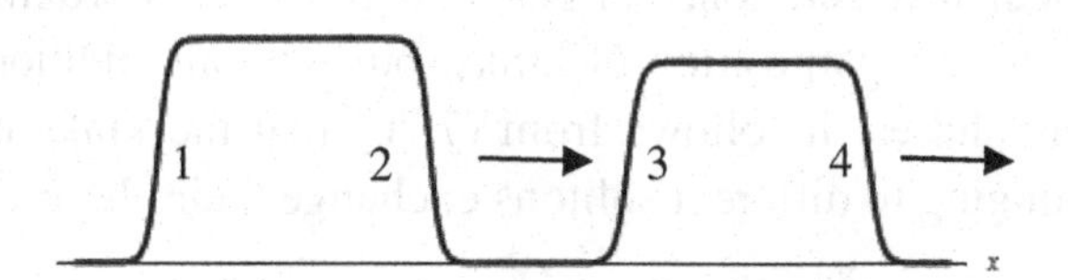

Figure 7.11. Scheme of interaction of two flat-top solitons.

Taking into account the Bäcklund transform in the form (7.64) with $D_1 = 0$, the solution of (7.69) for two solitons with the parameters $\mu_1 = 1 - k_1$ and $\mu_2 = 1 - k_2$ can be written in the form

$$S_{1,2} = \frac{1}{2}(\mu_1 + \mu_2)(X - 3T) \mp \Delta_+$$
$$-\ln\left\{\frac{4}{\mu_2 - \mu_1}\cosh\left[\frac{1}{2}(\mu_2 - \mu_1)(X - 3T \pm \Delta_-\right]\right\},$$
$$S_{3,4} = \frac{1}{2}(\mu_1 + \mu_2)(X - 3T) \mp \Delta_+ \tag{7.70}$$
$$+\ln\left\{\frac{4}{\mu_2 - \mu_1}\cosh\left[\frac{1}{2}(\mu_2 - \mu_1)(X - 3T \pm \Delta_-\right]\right\}.$$

Unknown parameters $\Delta_\pm$ defining distances between kinks can be found from the relations (7.62). At $D_1 = 0$ they are

$$\Delta_\pm = \Delta_1 \pm \Delta_2,\ 2\Delta_{1,2} = \ln(2/\mu_{1,2}). \tag{7.71}$$

Since all S_i depend on only one variable, $X - 3T$, the derivatives $\partial S / \partial T + 3\partial S / \partial X$ are zeros. Thus, the general solution which is the sum of the zero approximation (7.55) and the first approximation (7.66) at $N = 4$, acquires a simplified form (7.68):

$$u(x,t) = \frac{1}{2}\sum_{i=1}^{4}(-1)^{i+1}\left(1 - \frac{\partial S_i}{\partial X}\right)\tanh\frac{1}{2}\left[x - t - S_i(X,T)\right]. \tag{7.72}$$

As mentioned, this solution, corrected to the first approximation, still corresponds to a superposition of kinks, but with an additional variable shift of their phases. It follows from (7.70) that the kinks of the same polarity belonging to different solitons exchange their phases:

$$S_1(x,T \to \pm\infty) = S_3(x,T \to \mp\infty) + \ln\mu$$
$$= -(\Delta_+ \mp \Delta_-) \mp \mu(X - 3T) - \ln\mu,$$
$$S_2(x,T \to \pm\infty) = S_4(x,T \to \mp\infty) + \ln\mu \tag{7.73}$$
$$= (\Delta_+ \mp \Delta_-) \mp \mu(X - 3T) - \ln\mu.$$

Here $\mu = (\mu_1 - \mu_2)/2$. According to (7.73), after the interaction the solitons acquire the additional phase shifts, $\pm\ln(1/\mu)$, and this is the only final result of interaction, as is typical of the interaction of solitons as classical particles in integrable equations. However, these solutions also characterize solitons as waves. Indeed, the fact that e.g., a frontal kink of the leading soliton first affects the frontal kink belonging to the anterior soliton before interacting with the closer, rear kink, testifies to a distant action characteristic of fields rather than particles. This is illustrated in Fig. 7.12 which shows the trajectories of kinks, $X_i(t)$, which are defined by zeros of the hyperbolic tangents in (7.72):

$$x_i(t) = t + S_i\left(X_i - 3T\right),\ i = 1,...4. \tag{7.74}$$

It is seen from the figure that the frontal kink (S_4) begins to accelerate simultaneously with slowing down of the frontal kink (S_2) of the rear soliton; the same is true for the pair of rear kinks. This is due to the finite speed of transmitting energy from the fast to the slow soliton.

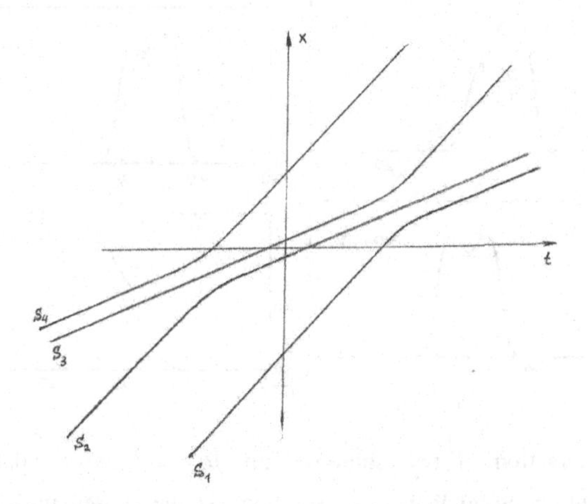

Figure 7.12. Typical trajectories of kink centers in two interacting flat-top solitons of the Gardner equation. Here $S_1 - S_2$ and $S_3 - S_4$ are the initial (at $t \to -\infty$) lengths of the fast and slow solitons, respectively.

This example allows a comparison with the known exact two-soliton solution of the Gardner equation [33, 34]. As shown in [17, 19], an exact N-soliton solution of the Gardner equation for an arbitrary number of solitons can be represented in a form similar to the approximate $2N$-kink solution (7.68), and the main terms of the corresponding expansions coincide. In particular, an exact two-soliton solution can be represented in a form analogous to (7.72) in which the phases differ by the terms of the order of μ^2.

We have already mentioned that the approximate approach developed here may be valid far beyond the limits in which the solitons have well-expressed, kink-like fronts and rears (according to Fig. 7.10, the latter is true only for very small μ). To illustrate this, in Fig. 7.13 we show the interaction of a flat-top soliton with a much smoother and smaller one, having $\mu = 0.4$ [17]. It is seen that even for such a large difference in the parameters of interacting solitons, an approximate solution remains reasonably close to the exact one, albeit with a slightly different rate of the process of overlapping and further separation of solitons.

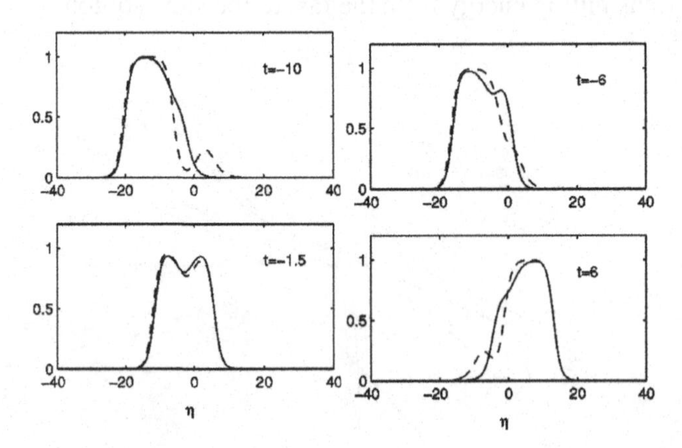

Figure 7.13. Interaction of two solitons with $\mu_1 = 10^{-6}$ and $\mu_2 = 0.4$. Dashed line: approximate solution. Solid line: exact solution. At the chosen initial conditions, the approximate and the exact solutions coincide at t = –1.5.

7.8.3. *Envelope solitons and kinks*

As shown above, the interaction of Gardner solitons is repulsive and can be described by the "double-Toda" system (7.65). As discussed above, for a lattice of Gardner solitons, perturbations can propagate as envelope waves in the form of Toda solitons and periodic envelope lattices; in turn, their perturbations also satisfy a Toda system, thus forming a lattice hierarchy. A specific feature of the present case is that the soliton is treated as a compound of two kinks and the lattices of frontal and trailing kinks are separated as described by two Toda subsystems (7.65).

Note first that Eqs (7.65) have two obvious solutions in the form of infinite periodic trains (lattices) of solitons:

$$S_n = \begin{cases} n\Lambda + Vt + \delta, & \text{for even } n, \\ n\Lambda + Vt \quad , & \text{for odd } n. \end{cases}, \tag{7.75}$$

This corresponds to periodic nonlinear waves in the original Gardner equation (7.49) having the spatial period 2Λ and the characteristic soliton length $2\delta < 2\Lambda$ (Fig. 7.14). Substitution of (7.75) into Eqs (7.65) yields the dispersion equation

$$3V = -2e^{-\Lambda}\cosh\delta, \tag{7.76}$$

which relates the speed of this periodic wave to the parameters Λ and δ.

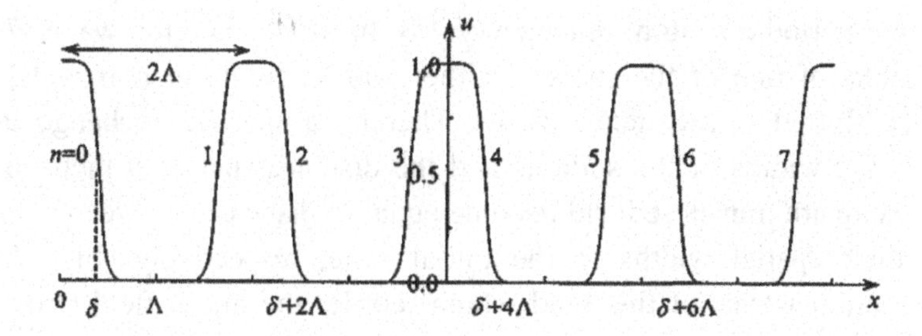

Figure 7.14. Sketch of a periodic chain of flat-top Gardner solitons. Positions of kinks and anti-kinks are indicated at the abscissa.

Modulation of these lattices is described by the Toda system (7.65) which, in turn, has solitary solutions that can be considered as pairs of kinks, and which are essentially the envelope solitons with respect to the basic solitons (7.51). A similar result was obtained above for soliton lattices in the KdV equation. In the present case, there exist two different types of localized structures that correspond to the excitation of Toda solitons either in one or in both subsystems. The respective steady-state solutions are

$$S_n(t)=\begin{cases} n\Lambda+Vt+\ln\left\{\dfrac{\cosh\left[\lambda(n-2)-\beta t\right]}{\cosh(\lambda n-\beta t)}\right\}, & \text{even } n, \\ n\Lambda+Vt, & \text{odd } n, \end{cases} \tag{7.77}$$

where $V=-2\cosh(2\lambda)\exp(-\Lambda)$, $\beta=\sinh(2\lambda)\exp(-\Lambda)$, $\delta=\pm 2\lambda$; and

$$S_n(t)=\begin{cases} n\Lambda+Vt+\delta+\ln\left\{\dfrac{\cosh\left[\lambda(n-2-d)-\beta t\right]}{\cosh\left[\lambda(n-d)-\beta t\right]}\right\}, & \text{even } n, \\ n\Lambda+Vt+\ln\left\{\dfrac{\cosh\left[\lambda(n-d)-\beta t\right]}{\cosh(\lambda n-\beta t)}\right\}, & \text{odd } n, \end{cases} \tag{7.78}$$

where V is the same as above but $\beta=\sinh(2\lambda)\exp(\Lambda)$ and $\sinh^2\lambda(1-d)=\sinh[\lambda^2(1+d)]\exp(2\delta)$. Note that in this latter case d and δ are independent parameters. Due to the symmetry of the subsystems, odd and even numbers for n in these expressions can be transposed.

These solutions describe modulation waves with respect to the original periodic soliton sequence (7.75). In the modulation wave (7.77) the kinks of one of the subsets (either soliton fronts or rear ends) are finally shifted at a distance of 4λ. Thereby a specific exchange takes place: the widths of the solitons and the distances between them in the final state are transposed and become equal to distances between solitons and their spatial widths in the initial state, respectively (Fig. 7.15). Modulation waves of this kind which reorder the initial field structure, can be called envelope kinks.

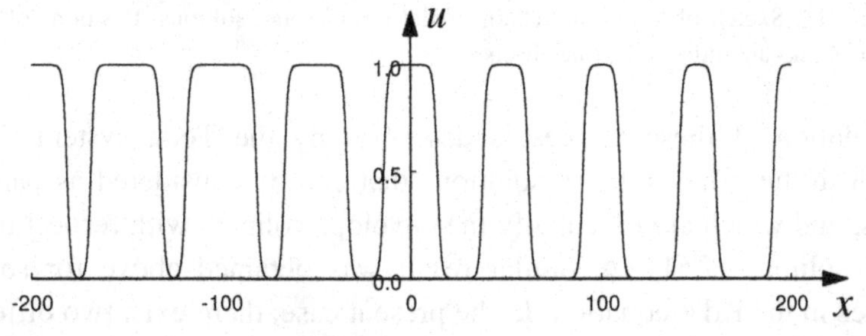

Figure 7.15. Example of the kink-like modulation of the sequence of Gardner solitons. Solitons and intervals between them are transposed from front to rear of the sequence. Numbers on the abscissa are arbitrary scales.

A modulation wave of the second kind, (7.78), shifts the kinks in both subsets (soliton fronts *and* rear ends) at the same distance 4λ. Therefore, wave profiles in the initial and final states are the same. Such modulation waves can be called envelope solitons (Fig. 7.16).

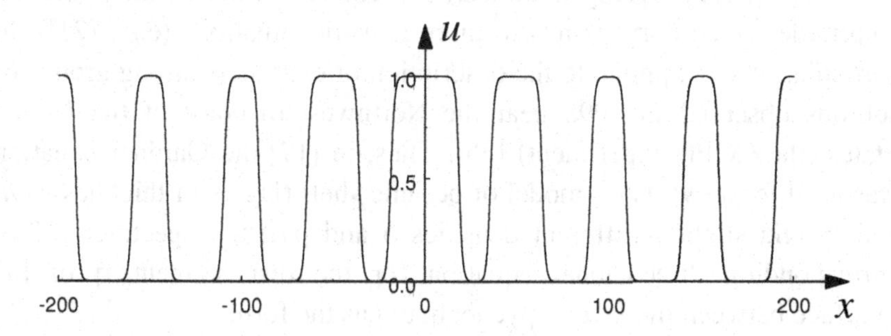

Figure 7.16. Example of the soliton-like modulation of the sequence of Gardner solitons. Solitons and intervals between them are transposed from front to center (at around $x = 0$) and then their order is restored. Numbers on the abscissa are arbitrary scales.

Envelope solitons and kinks can interact in the same way as in the basic Gardner equation. Moreover, due to their exponential asymptotics, approximate equations describing these interactions must again be Toda systems (7.65), which have solutions in the form of the "second-order" solitons and kinks. Continuing this process, a hierarchy of multi-periodic envelope waves can be constructed. Thus, the features of the considered solitons as compounds of kinks are reproduced at each level of such hierarchy.

7.8.4. *Physical example. Large-amplitude internal waves in the ocean*

The general scheme developed above was applied to the description of a sequence of internal solitons observed on the ocean shelf. In physical oceanography the main type of regularly observed solitary waves (or waves close to them) are internal gravity waves, generated by tides in the ocean shelf areas. During the past few decades it has been shown that such internal solitons are ubiquitous in many coastal zones (see the review paper [3] and references therein). More recently, it has been demonstrated that these waves are often strongly nonlinear and cannot be adequately described by the KdV equation used earlier to model such processes. Somewhat more adequate is the Gardner equation: although it is still essentially weakly nonlinear, it predicts a soliton with a limiting amplitude which corresponds to more realistic situations (e.g., [21]). In particular, it was applied to the description of extremely strong groups of solitons observed in 1995 near the Northwestern coast of the United States (the COPE experiment) [35]. Thus, in [17] the Gardner equation was used for a two-layer model of oceanic shelf (layers of thicknesses h_1 and h_2 and slightly different densities ρ and $\rho+\Delta\rho$, respectively). The corresponding dimensional equation for the displacement η of the interface between the layers (pycnocline) has the form

$$\frac{\partial \eta}{\partial t}+c\frac{\partial \eta}{\partial x}+\alpha\eta\frac{\partial \eta}{\partial x}+\alpha_1\eta^2\frac{\partial \eta}{\partial x}+\beta\frac{\partial^3 \eta}{\partial x^3}=0. \tag{7.79}$$

Here

$$c^2 = g\left(\Delta\rho/\rho\right)h_1h_2/\left(h_1+h_2\right);\ \alpha=\frac{3}{2}c\left(h_2-h_1\right)/h_1h_2;\ \beta=\frac{c}{6}h_1h_2;$$

$$\alpha_1=-\frac{3}{8}c\left(h_1^2+h_2^2+6h_1h_2\right)/h_1h_2,$$

and g is the gravity acceleration.

In the experiment, the displacement of the interface between the layers (the pycnocline) due to the internal waves was recorded in two locations: at 28 km from the coast (point A) and after its propagation for 20 km from this site, at 8 km from the coast (point B). To model this process, in [17] Eq. (7.79) was solved using the method described above, by approximating each soliton as a compound of two kinks. As an initial condition, a record of the wave in point A was used, and the wave in point B was calculated. Figure 7.17 shows the recorded wave (for the leading group of solitons) in point B together with the theoretical calculation.

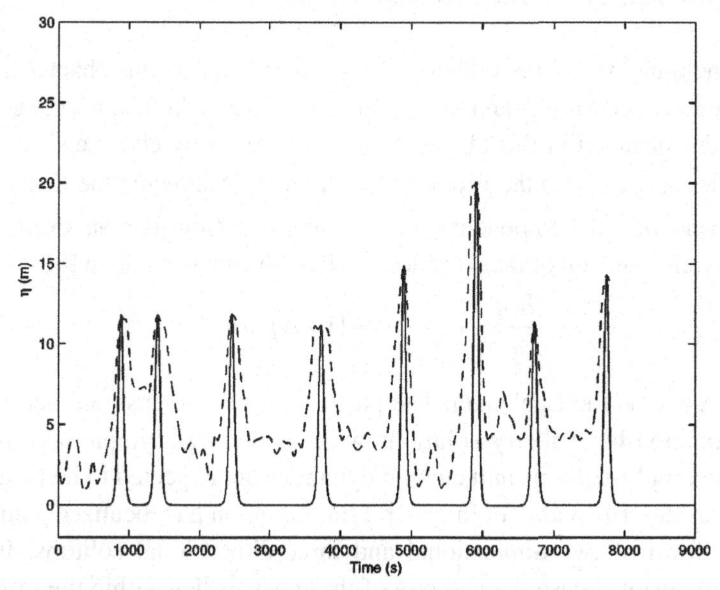

Figure 7.17. Calculated (solid line) and observed (dashed line) group of solitons at 8 km from the coast (point B), after the propagation at 20 km onshore from the point of the initial record (point A).

As seen from this figure, there are major similarities between the theory and observations in regard to the delays (velocities) and disposition of solitons on the time axis. Note that this behavior is radically different from that in the KdV case, in which solitons tend to acquire the amplitudes linearly increasing along the group. The main discrepancy between the theoretical model and the experiment is the difference in duration of impulses (2–3 mins in the experiment versus 1–1.5 mins in theory). This discrepancy is because, as mentioned, the Gardner equation is still adequate only when applied to moderately nonlinear processes. More recently, strongly nonlinear models have been suggested for a two-layer model. In [16] interaction of compound solitons by their kinks was studied for the so-called Choi–Camassa equations which are non-integrable and can define a larger length for solitons of the same amplitude.

7.9. Two- and Three-Dimensional Solitons

The one-dimensional perturbation scheme developed in this chapter can be readily extended to two- and three-dimensional cases. In fact, for the general equations obtained in this chapter, this can be done by changing the scalar variables x, ζ, X, to the vectors **r**, ζ, **X**, and by allowing the Lagrangian to depend on all components of these vectors. Consider an example: a conservative version of the so-called Swift–Hohenberg equation [8]:

$$\frac{\partial^2 u}{\partial t^2} + \alpha u - u^3 + (1+\Delta)^2 u = 0. \tag{7.80}$$

Here $\Delta = \nabla^2$ is the Laplacian. The particular, one-dimensional version of this equation has solitary solutions with oscillating asymptotics, so that complex multi-solitons and chaotic dynamics are expected to be possible, as it was described above. In general this equation has localized solutions in the form of two-dimensional and three-dimensional solitons. In the two-dimensional case the velocity of these waves lies within the limits of $0 \le V \le 0.9$, and their shape is an oblate oval. A particular case is an immovable soliton with $V = 0$. In this case a soliton becomes axially symmetric (Fig. 7.18).

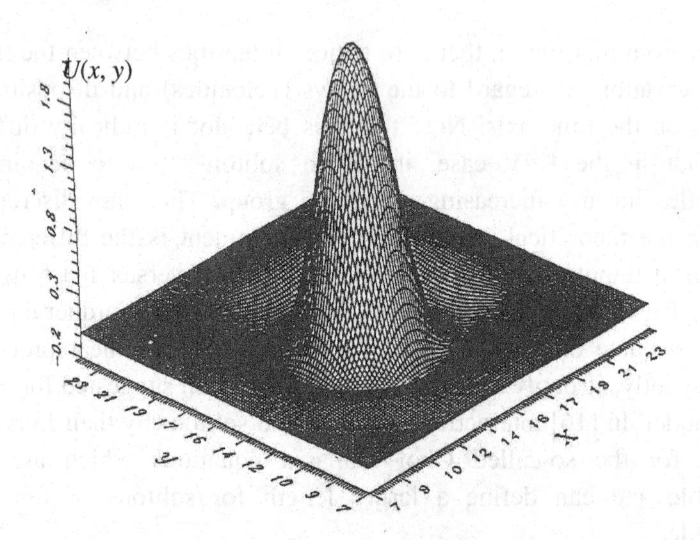

Figure 7.18. An immovable two-dimensional soliton in Eq. (7.80).

Its asymptotics which can, as before, be found by linearization of (7.80) (neglecting u^3), is proportional to the Bessel function $K_0(kr)$, where $k^2 = 1 + i\sqrt{\alpha}$, which has an oscillating structure.

Now consider the interaction of two such solitons [8]. Equation (7.80) can be represented in the Lagrangian form with the following Lagrangian density:

$$L = \frac{1}{2}u_t^2 + \frac{1}{2}(\nabla u)^2 - \frac{\alpha}{2}u^2 + \frac{1}{4}u^4 + \nabla u \nabla F + \frac{1}{2}F^2. \tag{7.81}$$

Here F is an auxiliary function such as $\nabla F \Delta u = 1 + 2\Delta u$.

According to the formulae (7.18) and (7.20) generalized to a two-dimensional case, for the mass of the equivalent classical particle and the pair interaction potential we have

$$m = \frac{\partial}{\partial V_x}\int U_t U_x d\zeta = \frac{\partial}{\partial V_y}\int U_t U_y d\zeta = \int_0^\infty (U_\zeta)^2 \zeta \, d\zeta,$$
$$W(S) = \int_0^\infty (U(\zeta))^2 u^{(1)}(|\zeta - \mathbf{S}|)\zeta \, d\zeta, \quad \zeta = |\zeta|. \tag{7.82}$$

Here the unperturbed solution U represents the symmetric soliton. According to (7.82), the problem of interaction of two solitons is reduced to that of particle motion in a central field:

$$\frac{d^2\mathbf{S}}{dt^2} = -\nabla W(S), \tag{7.83}$$

where **S** is the vector connecting soliton centers, and S is its modulus. As in the one-dimensional case, the structure of the pair potential $W(S)$ follows the dependence of the oscillating far field of the soliton.

Strictly speaking, the solitons moving due to the interaction cannot remain ideally symmetric. However, this asymmetry is small because the solitons' motions are slow due to the separation of soliton centers, and can be neglected in the main approximation.

As known in the classical mechanics [26], due to the angular symmetry, the problem (7.83) can be reduced to a one-dimensional one:

$$\frac{d^2S}{dt^2} = -\frac{d}{dS} W_{eff}(S),\ W_{eff} = W(S) + \frac{M^2}{S^2}, \tag{7.84}$$

where M is the angular momentum. The corresponding interaction potential and phase trajectories are shown in Fig. 7.19. This case is qualitatively similar to that shown in Fig. 7.2c, but now the motions are two-dimensional, and the vector **S** connecting the centers of the solitons can change its direction.

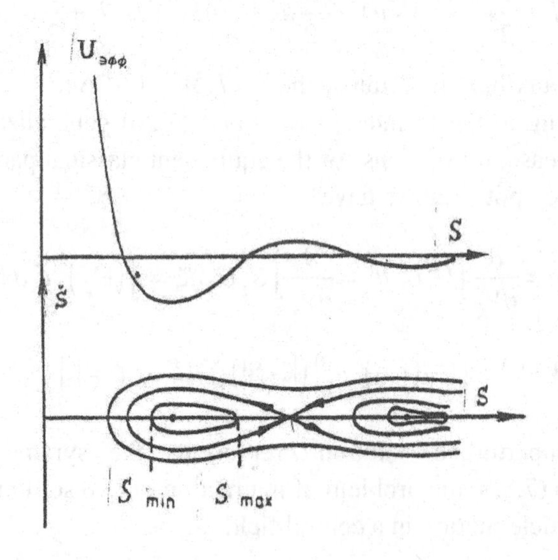

Figure 7.19. Interaction potential (top) and phase trajectories (bottom) for Eq. (7.84)

There exists a critical value M_{cr} such that if $M < M_{cr}$, the effective potential contains a finite number of minimums and maximums, each of which defines stable or unstable orbits of particles – solitons, rotating about their mass center. Near the stable equilibrium points (centers) on the phase plane shown in Fig. 7.19, the trajectories correspond to finite motions (they are not necessarily periodic, due to a possible direction variation). The trajectories close to the unstable equilibrium points (saddles) begin and end at infinity. They correspond to mutual scattering of solitons as particles according to (7.84). As in the classical mechanics [26], these trajectories can be characterized by the deviation angle χ defined as

$$\chi = \pi - 2\phi; \quad \phi = M \int_{S_{\min}}^{\infty} \frac{dS}{\sqrt{E - U_{эфф.}(S)}}. \tag{7.85}$$

Here E is the effective energy of the particle-soliton at the chosen trajectory. If the minimal distance $S_{\min}$ between the solitons approaches the radius of an unstable orbit, the angle χ becomes very sensitive to the initial scattering parameter M. In this case, spatial trajectories of particles have the form of spirals winding down and then (after reaching the circle $S_{\min}$) unwinding. This qualitative description is confirmed by direct numerical solution of Eq. (7.80). In particular, when M approaches its critical level M_{cr}, the motion becomes complex and eventually chaotic.

Of special interest is the existence of static solutions of (7.80) which satisfy the equation

$$\Delta^2 u + 2\Delta u + (\alpha + 1)u - u^3 = 0. \tag{7.86}$$

Evidently, this equation contains all one-dimensional multi-soliton solutions considered above for Eq. (7.46). In the two-dimensional and three-dimensional cases, stationary lattices (rectangular, hexagonal, etc.) are possible as well. Moreover, numerical solutions of similar equations demonstrate more complex bound states, such as tori, spirals, etc. (Fig. 7.18).

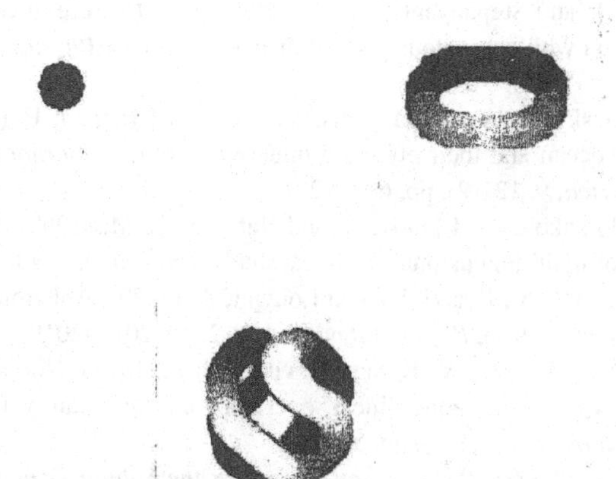

Figure 7.20. Examples of static localized structures in Eq. (7.86). From [5].

A more detailed analysis of these problems can be found in [7, 8].

7.9. Concluding Remarks

This chapter, already long enough, could be further extended by including other non-trivial cases of soliton interactions. Among them are, for example, the propagation of kinks in a smoothly inhomogeneous medium [20], the interaction of two-dimensional solitons in the Kadomtsev–Petviashvili (KP) equation [32], and the interaction of Rankine vortices and vortex pairs in ideal fluids [15]. Comparison with the available exact solutions shows that the approximate solutions considered here not only reflect the main qualitative features of soliton interactions but are often quantitatively valid beyond the formal limits of their applicability. The approximate approach is equally applicable for integrable and non-integrable equations, and even in the former case it typically yields much more graphic results than the exact methods.

References

1. Abramyan, L. A. and Stepanyants, Y. A. (1985a). Two-dimensional multisolitons: stationary solutions of the Kadomtsev–Petviashvili equation, *Radiophysics and Quantum Electronics*, v. 28(1), pp. 20–26.
2. Abramyan, L. A. and Stepanyants, Y. A. (1985b). The structure of two-dimensional solitons in media with anomalously small dispersion, *Soviet Physics JETP*, v. 61(5), pp. 963–966.
3. Apel, J., Ostrovsky, L. A., Stepanyants, Yu. A., and Lynch, J. F. (2007). Internal solitons in the ocean and their effect on underwater sound, *Journal of the Acoustic Society of America*, v. 121(2), pp. 695–722.
4. Aranson, I., Gorshkov, K., Lomov, A. and Rabinovich, M. (1990). Stable particle-like solutions of multidimensional nonlinear fields, *Physica D*, v. 43, pp. 435–453.
5. Bolchieri, P., Scotti, A., Bearzi, B., and Loinger, A. (1970). Anharmonic chain with Lennard-Jones interaction, *Physical Review A*, v. 2, pp. 2013–2019.
6. Dubrovin B. A., Matveev, V. B., and Novikov, S. P. (1976). Non-linear equations of the KdV type, finite zone linear operators and Abelian varieties, *Russian Mathematical Surveys*, v. 31(1), pp. 59–146.
7. Gorshkov, K. A. (2007). *Perturbation Theory in the Soliton Dynamics*, Doctor of Science Thesis, Institute of Applied Physics, Nizhny Novgorod. [in Russian].
8. Gorshkov, K. A., Lomov, A. S., and Rabinovich, M. I. (1992). Chaotic scattering of two-dimensional solitons, *Nonlinearity,* v. 5, pp. 1343–1353.

9. Gorshkov, K. A. and Ostrovsky, L. A. (1981). Interactions of solitons in non-integrable systems: Direct perturbation method and applications. *Physica D*, v. 3, pp. 428–438.
10. Gorshkov, K. A. and Ostrovsky, L. A. (1984). Interaction of solitons with their own radiation fields, *3rd International Symposium on selected problems of statistical mechanics, Dubna, Russia*, v. 2, 222–226 [in Russian].
11. Gorshkov, K. A., Ostrovsky, L. A., and Papko V. V. (1976). Interactions and bound states of solitons as classical particles, *Soviet Physics JETP*, v. 44, pp. 306–311.
12. Gorshkov, K. A., Ostrovsky, L. A., and Papko V. V. (1977). Soliton turbulence in a system with weak dispersion, *Soviet Physics Doklady*, v. 22(7), pp. 378–380.
13. Gorshkov, K. A., Ostrovsky, L. A., Papko, V. V. and Pikovsky, A. S. (1979). On the existence of stationary multisolitons, *Physics Letters A*, v. 74, pp. 177–179.
14. Gorshkov, K. A. and Papko V. V. (1977). Dynamic and stochastic oscillations of soliton lattices, *Soviet Physics JETP*, v. 46, pp. 92–97.
15. Gorshkov, K. A, Ostrovsky, L.A., and Soustova, I. A. (2000). Perturbation theory for Rankine vortices, *Journal of Fluid Mechanics*, v. 404, pp. 1–25.
16. Gorshkov, K. A, Ostrovsky, L.A. and Soustova, I. A. (2011). Dynamics of Strongly Nonlinear Kinks and Solitons in a Two-Layer Fluid, *Studies in Applied Mathematics*, v. 126, pp. 49–73.
17. Gorshkov, K. A, Ostrovsky, L.A., Soustova, I.A., and Irisov, V. G. (2004). Perturbation theory for kinks and its application for multi-soliton interactions in hydrodynamics, *Physics Review E*, v. 69: 016614, pp. 1–9.
18. Gorshkov, K. A., Ostrovsky, L. A., and Stepanyants, Y. A. (2010). Dynamics of soliton chains: From simple to complex and chaotic motions, *Long-Range Interactions, Stochasticity and Fractional Dynamics*, eds. A. Luo and V. Afraimovich, Springer, London and New York, pp. 177–218.
19. Gorshkov, K. A and Soustova I. A. (2001). Interaction of solitons as compound structures in the Gardner model, *Radiophysics and Quantum Electronics*, v. 44, pp. 465–476.
20. Gorshkov, K. A., Soustova, I. A., Ermoshkin, A. V., and Zaytseva, N. V. (2012). Evolution of the compound Gardner-equation soliton in media with variable parameters, *Radiophysics and Quantum Electronics*, v. 55, pp. 344–356.
21. Grimshaw. R, Pelinovsky, D., and Talipova, T. (1997). The modified Korteweg–de Vries equation in the theory of large-amplitude internal waves, *Nonlinear Processes in Geophysics*, v. 4, pp. 237–250.
22. Kakutani, T. and Ono, H. (1969). Weak nonlinear hydromagnetic waves in a cold collision-free plasma, *Journal of the Physical Society of Japan*, v. 26, pp. 1305–1318.
23. Kawahara, T (1972). Oscillatory solitary waves in dispersive media, *Journal of the Physical Society of Japan*, v. 33, pp. 260–264.
24. Kuznetsov, E. A and Mikhailov, A. V, (1974). Stability of stationary waves in nonlinear weakly dispersive media, *Soviet Physics JETP*, v. 40(5), pp. 855–859.
25. Lamb, G. L. (1980). *Elements of Soliton Theory*. J. Wiley & Sons, New York.
26. Landau, L. D. and Lifshits, E. M. (1976). *Mechanics*. Elsevier, Oxford.

27. Lighthill, M. J. (1965). Contributions to the theory of waves in nonlinear dispersive systems, *Journal of the Institute of Mathematical Applications*, v. 1, pp. 269–306.
28. Manakov, S. V. (1974). Complete integrability and stochastization of discrete dynamical systems, *Soviet Physics JETP*, v. 40, pp. 269–274.
29. Matveev, V. B. and Salle, M. A. (1991). *Darboux transformation and solitons.* Springer, Berlin.
30. Nagashima, H. (1979). Experiment on solitary waves in the non-linear transmission-line described by the equation $du/dt + udu/dz = d^5u/dz^5 = 0$, *Journal of the Physical Society of Japan*, v. 47, pp. 1387–1388.
31. Obregon, M. A. and Stepanyants, Y. A. (1998). Oblique magneto-acoustic solitons in a rotating plasma. *Physical Letters. A*, v. 249, pp. 315–323.
32. Pelinovsky, D. E. and Stepanyants, Y. A. (1993). New multisoliton solutions of the Kadomtsev–Petviashvili equation, *JETP Letters*, v. 57(1), pp. 24–28.
33. Slyunyaev A. V. (2001). Dynamics of localized waves with large amplitude in a weakly dispersive medium with a quadratic and positive cubic nonlinearity, *JETP*, v. 92, pp. 529–534.
34. Slyunyaev. A. V. and Pelinovsky, E. N. (1999). Dynamics of large-amplitude solitons, *JETP*, v. 89, pp. 173–181.
35. Stanton, T. P. and Ostrovsky, L. A. (1998). Observations of highly nonlinear internal solitons over the Continental Shelf, *Geophysical Research Letters*, v. 25, pp. 2695–2698.
36. Stepanyants, Y. A. (2005). Dispersion of long gravity-capillary surface waves and asymptotic equations for solitons, *Proceedings of the Russian Academy of Engineering Sciences, Series Applied Mathematics and Mechanics*, v. 14, pp. 33–40 [in Russian].
37. Toda, M. (1989). *Theory of Nonlinear Lattices*, 2nd ed. Springer, Berlin.
38. Zabusky, N. J. and Kruskal, M. D. (1965). Interaction of "solitons" in a collisionless plasma and the recurrence of initial states, *Physics Review Letters*, v. 15(6), pp. 240–243.

Chapter 8

Dissipative and Active Systems. Autowaves

Never express yourself more clearly than you think.

N. Bohr

The problems considered in this final chapter are specific in two aspects. First, here we concentrate on dissipative and especially on active processes, called *autowaves*, a term that was apparently introduced by R. Khokhlov (see [18]). They exist when there is a source of energy which, being balanced by dissipation, leads to the establishment of a steady-state regime (as was shown in Chapter 1 for a limit cycle). Second, in most examples considered here, the process consists of "slow" and "fast" phases, strongly differing by their scales in time and/or space. Such waves are encountered in an extremely wide range of physical and biological studies; here we limit ourselves to a few characteristic models.

8.1. Burgers Equation and Taylor Shocks

We begin from a relatively simple nonlinear equation which includes dissipation:

$$u_t + uu_x - \mu g u_{xx} = 0. \tag{8.1}$$

This is the *Burgers equation* which combines nonlinearity and diffusion in a simple way. It has a rather long history beginning in the early 1900s; it was named after J. Burgers after he used it in 1948 as a simple model of hydrodynamic turbulence. In the early 1950s, E. Hopf and J. Kole showed that the general solution of this equation can be obtained using the change of variables: $u = 2\mu g s_x / s$; after that it is reduced to the classic linear diffusion equation: $s_t = \mu g s_{xx}$ for which a general solution can be written in the form of a Fourier integral; this is the so-called Hopf–Cole transform (note that it was independently found by A. Forsyth in 1906 and by V. Florin in 1948). This solution, as well as elements of the history of the Burgers equation can be found in the books [16, 19]. The Burgers equation was presumably the first physically sound integrable nonlinear PDE.

Even after formal integration of this equation in terms of $s(x,t)$, returning to the initial physical variables which satisfy a given initial condition can be a problem in itself. On the other hand, in a variety of applications, the case of $\mu \ll 1$ is most interesting. In this case, an approximate approach based on matching the "fast" and "slow" solutions can be used. The slow part of the process corresponds to $\mu = 0$. The remaining first-order equation has an implicit solution in the form of the simple wave:

$$u = F(X - uT), \text{ or } X = uT + \Phi(u). \tag{8.2}$$

Here, as above, $X = \mu x$, $T = \mu t$, F is an arbitrary function defined by the initial condition, and Φ is the function inverse to F. We already briefly discussed such solutions in Chapter 4 where they described waves of modulation. In Eq. (8.1), u is a primary physical variable such as, for example, pressure in an acoustic wave. In this wave each fixed level of u propagates at a constant velocity equal to u (as mentioned above for the KdV equation, this can be a small addition to the linear wave velocity c_0). In general, the wave front is steepening. As was discussed earlier in this book, after a finite time the wave profile becomes two- or three-valued, so that the term with u_{xx} cannot be neglected anymore, and (8.2) must be matched with the "fast" solution involving all terms in (8.1). This matching is similar to that used in the boundary layer theory. The fast part can be sought as a solution of the stationary, ODE version of (8.1) describing solutions depending on $\zeta = x - Vt$. The latter is the known Taylor shock wave which is a transition between two constant asymptotics, U_1 and U_2 at $\zeta \to \pm\infty$. In the case of $U_1 = 0$ considered below (this does not significantly limit the generality) this solution is

$$u = U = \frac{U_2}{2}\left[1 - \tanh\left(\frac{U_2\zeta}{4\mu g}\right)\right], \; V = \frac{U_2}{2}. \tag{8.3}$$

To match the solutions (8.2) and (8.3) one has to suppose that the asymptotics $U_2 = U_2(T)$ and, consecutively, the shock velocity $V(T)$, slowly vary in time.

Now one can construct a solution consisting of a shock followed by a simple wave (Fig. 8.1). The matching of these two solutions occurs at the coordinate $X = X_s(T)$ of the fast solution which, in the first approximation, can be treated as a thin, quasi-stationary shock front.

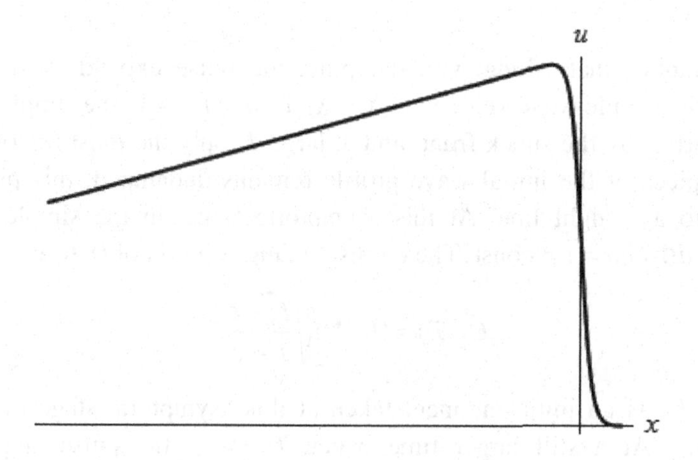

Figure 8.1. Schematic plot of a shock wave followed by a long simple wave.

According to (8.3),

$$\frac{dX_s}{dT} = V(T) = \frac{U_2(T)}{2}. \tag{8.4}$$

At the same time, from (8.2) it follows that

$$\frac{dX_s}{dT} = U_2 + \left(T\frac{du}{dT} + \Phi_u \frac{du}{dT}\right)_{u=U_2}. \tag{8.5}$$

From (8.4) and (8.5) we have

$$\frac{dU_2}{dT}+\frac{1}{2}\left(T+\frac{d\Phi}{du}\right)^{-1}_{u=U_2}U_2=0. \tag{8.6}$$

For the stretching part shown in Fig. 8.1 (in gas dynamics it is often called a "rarefaction" wave), $d\Phi/du>0$. Hence, the shock amplitude U_2 decreases in the course of propagation, which is natural in the presence of dissipation. From (8.3) it follows that the shock velocity V also decreases. Note that integrating (8.1) over x for a finite-length impulse, we have the "momentum conservation":

$$M=\int_{-\infty}^{\infty}udx=\text{const.} \tag{8.7}$$

This implies that, along with damping, the pulse expands with time. Since the simple wave velocity at $X = X_s$ is $u = U_2 > V$, the simple wave is absorbed by the shock front, and at large T, only the most remote and small piece of the initial wave profile remains undamped; this piece is close to a straight line. At this asymptotic stage, in the simple wave "tail," $d\Phi/du=\tau=\text{const}$. The corresponding solution of (8.6) is

$$U_2(T)=U_2(T_0)\sqrt{\frac{T_0+\tau}{T+\tau}}, \tag{8.8}$$

where T_0 is an initial moment taken at this asymptotic stage of wave damping. At a still larger time, when $T >> \tau$, the pulse amplitude decreases as $T^{-1/2}$.

Let us compare variations of the spatial scale of the shock front, L_1, which, according to (8.3), has the order of $4\mu g/U_2$, and the scale of the simple wave "tail", $L_2 = X_s - X_0$, where X_0 is the position of the rear point where $u = 0$ (note that this point is fixed; without loss of generality, we can let $X_0 = 0$.) The ratio of these two scales is

$$\frac{L_2}{L_1}\sim\frac{X_sU_2}{4\mu g}=\text{const.} \tag{8.9}$$

Indeed, after substituting (8.8) into (8.4), we find that X_2 increases as $\sqrt{q+T}\propto 1/U_2$. Thus, during its damping, the pulse preserves a proportion between the lengths of its front and "tail".

Note finally that at the asymptotic stage, the solution of the Burgers equation is known as the self-similar wave depending on the ratio $x/\sqrt{t}$. In the most interesting case when the ratio (8.9) is large, the solution has the form

$$u = \sqrt{\frac{g}{t}} \frac{e^{R(1-z^2)}}{\sqrt{\pi} + e^R \int\limits_{z\sqrt{R}}^{\infty} e^{-y^2} dy}, \tag{8.10}$$

where $z = x/\sqrt{2Mt}$, M is a constant momentum defined by (8.7), and the parameter $R = M/2g >> 1$ can be considered as the wave's Reynolds number. The corresponding wave profile is shown in Fig. 8.2.

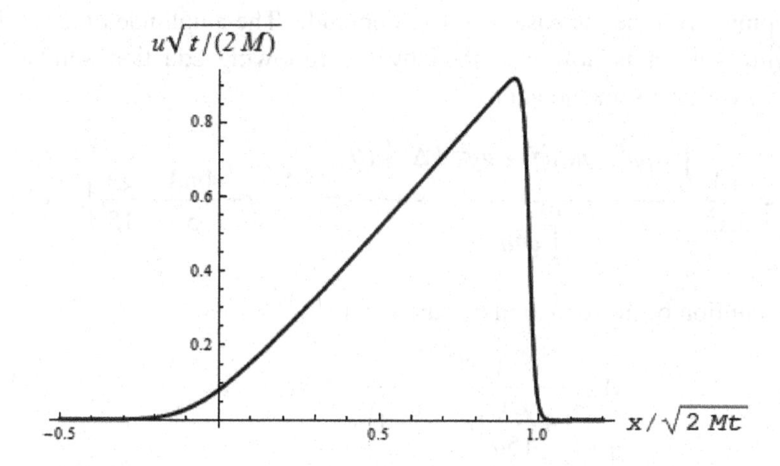

Figure 8.2. Self-similar solution (8.10) at $R = 40$ in dimensionless variables.

It is easy to see that at $\mu << 1$ this solution possesses the properties described above: it can be obtained by matching of the shock front with the simple "rarefaction" wave.

Shock waves and simple waves are typical objects of compressible fluid dynamics [1], nonlinear acoustics [12], electrodynamics of nonlinear media [4], and many other areas of physics.

8.2. Autosolitons and Explosive Instability

In Chapter 6 we analyzed the KdV equation (6.14) perturbed by dissipation. Consider now an equation similar to (6.14) with different signs of terms with q and m which now describe "negative losses":

$$u_t + uu_x + u_{xxx} = \mu(qu + mu^3 + gu_{xx}). \tag{8.11}$$

Here the parameters g, m, and q are assumed positive. Now the term with g describes dissipation, as before, whereas the parameter q can be positive only in the presence of a source able to "pump" energy into the system. Note that now the nonlinear term with m is not associated with friction; it describes the possible growth of negative losses (intensification of pumping) with the increase of wave amplitude. The amplitude of a slowly varying soliton is now described by the following equation similar to (6.15), with the same notations:

$$\frac{dA}{dT} = \frac{4A}{3\Delta^2} \frac{\int_{-\infty}^{\infty} \left[q\varphi^2 + mA\varphi^3 - g\varphi_\theta^2 / \Delta^2\right] d\theta}{\int_{-\infty}^{\infty} \varphi^2 d\theta} = \frac{4A}{3}\left(q + \frac{4mA}{5} - \frac{gA}{15}\right). \tag{8.12}$$

The solution of this equation has the form

$$A = \frac{A_0 e^{4qT/3}}{1 + \frac{\gamma A_0}{15q}\left(e^{4qT/3} - 1\right)}, \quad \gamma = g - 12m. \tag{8.13}$$

At the initial stage, when A_0 is sufficiently small so that the term with γ in the denominator is negligible, the soliton amplitude increases exponentially; in other words, the equilibrium state $A = 0$ is unstable. The further dynamics depends on the sign of γ. If $\gamma > 0$, the growth slows down, and the soliton reaches another equilibrium state with a constant amplitude $A = A_s = 15q/\gamma$. This state is stable, i.e., small deviations of amplitude from A_s eventually disappear. In the stable equilibrium state, dissipation due to the viscous term with γ is balanced by the energy pumping due to the "active" term with q. Such a wave can be called *autosoliton*. This process is shown in Fig. 8.3.

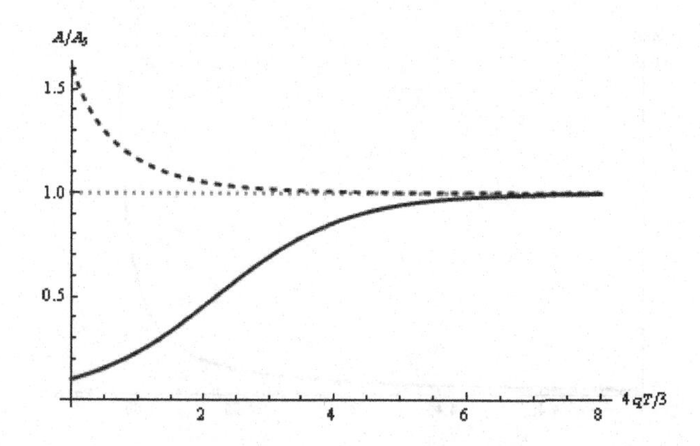

Figure 8.3. Soliton amplitude versus normalized time in the model (8.11) of an active/dissipative medium (solid lines 1 and 2) at $\gamma > 0$ and two initial conditions: $A_0 = 0.1\ A_s$ (solid line) and $A_0 = 1.6\ A_s$ (dashed line).

Quite different is soliton behavior when $\gamma < 0$, i. e., $m/g > 1/12$. In this case the soliton amplitude diverges at a finite time interval. This is an example of *explosive instability*. The time of this "explosion" is

$$T_{ex} = \frac{3}{4q} \ln\left(1 + \frac{15q}{|\gamma| A_0}\right). \tag{8.14}$$

The corresponding process is shown in Fig. 8.4.

An important difference from the dissipative case considered in Chapter 6 is that the soliton width now decreases. It is easy to see that each term in the left-hand side of (8.11) increases as $A^{5/2}$, whereas the terms in the right-hand side do not increase faster than A^2. As a result, if the initial condition secures the smallness of the terms in the right-hand side, in the course of wave amplification, these terms will remain relatively small and, unlike in the dissipative case, within the framework of (8.11) the solution (8.13) never fails. This, however, may not be true in a real physical system because the KdV equation is actually valid at small nonlinearity which ceases to be small in the course of soliton amplification, particularly in the explosive case. For example, amplification of a water wave due to wind or a shear current can result in the wave breaking.

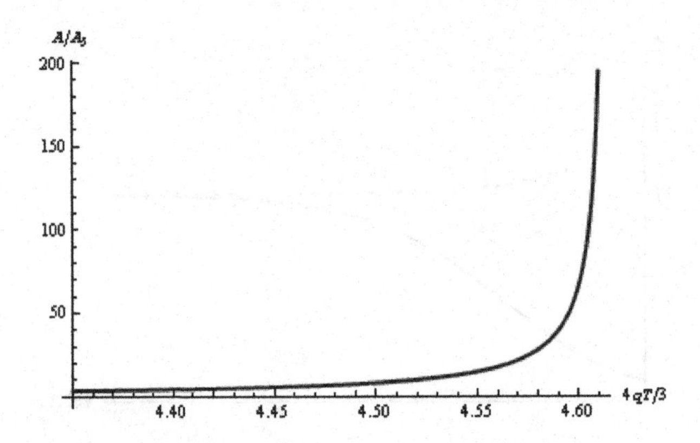

Figure 8.4. Soliton amplitude growth near the "explosion time" ($4qT_{ex}/3 \approx 4.615$) versus normalized time in the solution (8.13) at $\gamma < 0$ and $A_0 = 0.01A_s$, where A_s corresponds to the equilibrium amplitude at $\gamma > 0$.

8.3. Parametric Amplification of Solitons

8.3.1. *Solitons as accelerated particles*

When considering the interaction of solitons in Chapter 7 we concentrated on solitons having close amplitudes (more exactly, velocities) when their parameters vary relatively slowly. Here we consider the propagation of a soliton on the background of a small but long wave which can change its velocity and energy. In the KdV equation

$$u_t + uu_x + u_{xxx} = 0 \tag{8.15}$$

we seek the solution in the form $u = u_1 + u_2$, where u_1 is a given, long-wave solution of (8.15) and u_2 is close to a soliton with variable parameters, which satisfies the equation that follows from (8.15) after subtraction of the same equation for u_1:

$$u_{2t} + u_2 u_{2x} + u_{2xxx} + (u_1 u_2)_x = 0. \tag{8.16}$$

Note that there exists a trivial solution $u_2 = 0$, and u_1 enters not as an external force but as a slowly varying parameter. We suppose that u_1 is a small-amplitude harmonic wave:

$$u_1 = \mu a \cos(kX - \Omega T). \tag{8.17}$$

The solution for u_2 is represented similarly to (6.3):

$$u_2 = A(T)\text{sech}^2\left[\sqrt{\frac{A(T)}{12}}(\zeta - S(T))\right] + \mu u_2^{(1)} + \dots \tag{8.18}$$

Here S characterizes the position (phase) of the soliton center with respect to the closest maximum of the wave.

First, it is easy to write a "kinematic" equation describing variation of the phase S. It is defined by the difference between the velocities of the parameter wave u_1 and the local soliton velocity, $V + \mu a\cos(kX - \Omega T)$; the addition is due to the term $\mu u_2 u_{1x}$ in (8.16). Hence,

$$\frac{d\varphi}{dT} = k\left(\frac{A}{3} + a\cos\varphi - V_p\right), \tag{8.19}$$

where $\varphi = kS$ is the dimensionless phase and $V_p = \Omega / k$ is the velocity of the pump wave (8.17).

The equation for soliton amplitude can be obtained in the framework of the perturbation scheme described in Chapter 6. More specifically, we use Eq. (6.11) (in which the blind variable φ is evidently not the same as $\varphi(T)$ here) and substitute $R = -(u_1 u_2)_x$, to obtain

$$\frac{dA}{dT} = \frac{4}{3}ak\sin\varphi. \tag{8.20}$$

Equations (8.19) and (8.20) define the variation of soliton amplitude and its position on the pump wave. Now generalize these equations to a dissipative equation such as Eq. (6.14) with $m = 0$:

$$u_t + uu_x + u_{xxx} = -\mu(-gu_{xx} + qu). \tag{8.21}$$

In this case Eq. (8.19) remains unchanged, whereas (8.20) becomes

$$\frac{dA}{dT} = \frac{4}{3}\left[(ak\sin\varphi - q)A - \frac{g}{15}A^2\right]. \tag{8.22}$$

Similar equations (obtained in a different way) were discussed in the papers [6, 5] and in the book [14]. Here we describe the main features of their solutions and then discuss some generalizations.

First, define the equilibrium states of the system (8.19) and (8.22) at which $dA/dT = d\varphi/dT = 0$. One pair of the corresponding stationary solutions is trivial:

$$A_{3,4} = 0, \ \cos\varphi_{3,4} = V_p / a. \tag{8.23}$$

These equilibria exist if $a \geq |V_p|$. The other non-zero equilibrium states are

$$A_{1,2} = 3\frac{k^2V_p - 3bq \pm k\sqrt{(k^2+9b^2)a^2 - (q+3bV_p)^2}}{k^2+9b^2},$$
$$\cos\varphi_{1,2} = \frac{3b(q+3bV_p) \mp k\sqrt{(k^2+9b^2)a^2 - (q+3bV_p)^2}}{a\left(k^2+9b^2\right)}. \tag{8.24}$$

Here $b = g/15$. These non-zero equilibria exist only at a sufficiently large a, namely,

$$a > a_{cr} = \frac{q+3bV_p}{\sqrt{k^2+9b^2}}. \tag{8.25}$$

Note also that only $A_{3,4} > 0$ are physically sound values; indeed, in Eq. (8.15) solitons are always positive.

In [6] this process was analyzed in the case of $b = 0$, $q \neq 0$. The result qualitatively depends on the relation between the parameters a, V_p, and q. Figure 8.5 shows the phase plane of variables (φ, A) for the system (8.19), (8.22) in different regimes. These patterns are 2π - periodic in φ.

In the case (a) the trivial state $A = 0$ is unstable, and a small initial perturbation grows up to the non-zero equilibrium value A_2. In the case (b), small perturbations attenuate to zero amplitude and equilibrium phases φ_3 or φ_4, whereas the larger solitons with a proper initial phase can acquire the non-zero stationary amplitude A_2. The case (c) is similar to (b) but there is no definite equilibrium phase for small solitons.

Note that in this example the soliton again behaves as a particle which can be accelerated. Indeed, Eqs (8.19) and (8.22) are similar to those used in simple modeling of cyclic accelerators of particles [11].

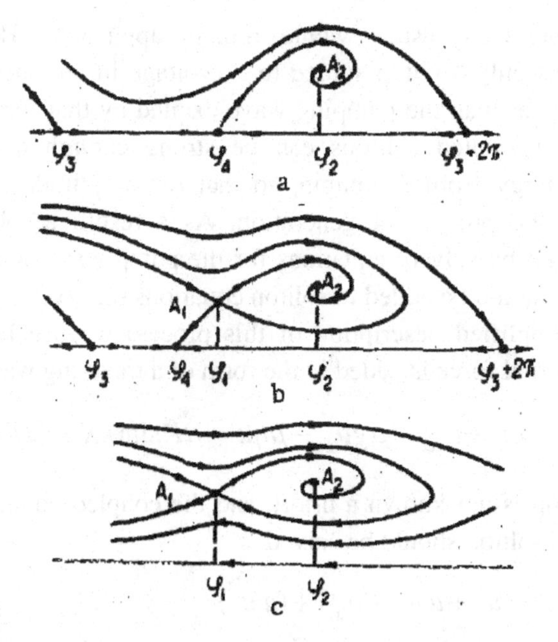

Figure 8.5. Phase trajectories of system (8.19), (8.22) on the plane (φ, A) for three different values of velocity detuning V.

8.3.2. *Parametric generation of solitons*

As seen from Fig. 8.5, in the first regime (a) an initial soliton with a small amplitude (small but still sufficient to consider it much narrower than the scale of the long, "pump" wave) increases until it reaches a non-zero equilibrium state. This process can be considered as parametric amplification of a soliton. The effect can be enhanced if a soliton is excited in a bounded system, a resonator, which supports many cycles of interaction. It can be a "ring" resonator in which a soliton is moving uni-directionally, or a resonator with reflecting ends. In the latter, the pump wave is a standing wave, but the soliton resonantly interacts with one of the traveling components of the standing wave propagating in the same direction as the soliton, whereas its interaction with the oppositely propagating wave is non-resonant and it results only in small variations of its amplitude and velocity. Hence, if the pump wave in a resonator is given, the previous results would remain applicable. However, in practical cases only a pump source (e.g., voltage in an electromagnetic resonator), rather than the pumping wave excited by that force, is given. Moreover, the excited solitons can be strong enough to take away significant energy from the pump, so that the amplitude of the latter decreases in the process of generation. As a result, a self-consistent problem should be solved: equations for the pump wave amplitude a in the resonator should be added to soliton equations (8.19) and (8.22).

For a simplified description of this process we use Eq. (8.21) in which an external force is added in the form of a traveling wave:

$$u_t + uu_x + u_{xxx} + \mu g u_{xx} - \mu q u = \mu P \sin(kX - \Omega T). \qquad (8.26)$$

Now the pump is not known a priori, and the coupled equations for the pump and the soliton should be solved:

$$\begin{aligned} &u_{it} + u_i u_{ix} + u_{ixxx} + (u_i u_j)_x \\ &= \mu\{g u_{ixx} - q u_i + P\cos(kX - \Omega T)\}, \end{aligned} \qquad (8.27)$$

where i,j = 1,2. Since the term with P is slowly varying, the above equations for the soliton, i = 2, remain unchanged. The pump wave u_1 in the linear approximation is represented by (8.17); however, now its amplitude, $a = a(T)$, is to be found. As soon as nonlinearity for u_1 is neglected, we can use the method used in Chapter 2 for the quasi-harmonic wave (8.17), to obtain

$$\frac{da}{dT} = -a(q + gk^2 + F(A,k)\sin\varphi) + P. \qquad (8.28)$$

Here the term with F is the amplitude of the first Fourier harmonic of the last term in the left-hand side of (8.27) at i = 2. Depending on the phase φ, a soliton can receive energy from the pump or release energy to it.

Now the system of three equations, (8.19), (8.22), and (8.28) should be analyzed. If the effect of a soliton on the pump can be neglected (F = 0), in the stationary regime $a = P/(q + gk^2)$, and the stationary soliton amplitudes in (8.24) can be rewritten in terms of the external forcing P:

$$A_{3,4} = 3P \frac{k^2 V_p - 3gq \pm \sqrt{k^2 \left[P^2 \dfrac{\left(9g^2 + k^2\right)}{\left(q + gk^2\right)^2} - \left(q + 3gV_p\right)^2 \right]}}{9g^2 + k^2}. \tag{8.29}$$

Thus, the non-zero equilibrium states exist if

$$P > \frac{\left(q + gk^2\right)\left(q + 3gV_p\right)}{\sqrt{k^2 + 9g^2}}. \tag{8.30}$$

With F taken into account, the analytical solution is cumbersome. Since the soliton width is much smaller than the pump's half-wavelength, i.e., $A / k^2 >> 1$, the function F can be expressed in hypergeometric functions. However, in real experiments with resonators, the problem is even more complicated. First, the excitation of the pump typically occurs at a fixed point of a ring resonator or at a boundary of a finite-length resonator with reflecting ends, rather than along the whole system. Second, the system can be excited at a higher-order mode when several half-waves of the pump exist in the resonator. Finally, there could be several solitons at each half-wavelength of the pump or, conversely, not all periods of the pump can support a soliton. These problems were considered in [6], where the corresponding theory was developed and soliton generation was experimentally observed in different regimes. As a result, instead of (8.28), two equations were obtained; at $g = 0$ they are:

$$\begin{aligned} &\left[kL\frac{da}{dT} + qa + \sqrt{6}n_s A^{3/2} \right] \cos \pi n(\Omega / \Omega_n) = P\cos\Theta, \\ &kLa\frac{d\Theta}{dT}\cos \pi n(\Omega / \Omega_r) + a\sin \pi n(\Omega / \Omega_n) = -P\sin\Theta. \end{aligned} \tag{8.31}$$

Here L is the length of the resonator, Θ is the phase difference between the pump wave in the resonator and the pump source at the excitation point, n_s is the total number of solitons in the resonator, n is the mode number in the resonator, and Ω_n is the resonance frequency of that mode. In [6] these equations were analyzed and the solutions compared with the experimental data obtained in a nonlinear electromagnetic line. Figure

8.6 shows a few of the experimentally observed wave shapes. In the last two photos doubling of the period of soliton sequence is clearly seen.

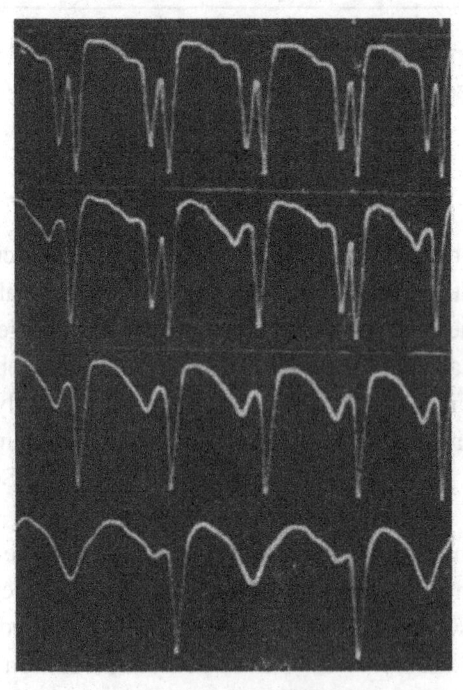

Figure 8.6. Records of oscillations in a resonator excited by a harmonic pump source at a frequency close to that of the second mode of the resonator (n = 2) at different small detunings from the exact linear frequency of the second mode.

8.3.3. *Two-dimensional resonators*

The above consideration can be extended to two-dimensional resonators. It was done in [15] based on the approximation in which the interaction between solitons propagating in different directions is neglected, because in the integrable systems such solitons quickly diverge and preserve their parameters, and if the soliton is much shorter than its path between the reflections from the resonator boundaries, its interactions with re-reflected soliton fronts are unimportant. On the other hand, this "billiard" of soliton fronts should be resonant, i.e., the reflection process should be periodic. For plane fronts in an isotropic system it is possible in four geometries: a rectangle; a right-angle isosceles triangle; an equilateral triangle; and a right-angle triangle with acute angles of $\pi/6$ and $\pi/3$. However, the experiments were carried out in a discrete, two-dimensional electric lattice in which dispersion occurs due to the discreteness. Such systems are described by differential-difference equations. For long waves the differences can be expanded to obtain differential wave equations. In particular, for each plane front, the KdV equation can be written in which the dispersion term depends on the propagation direction angle ψ with respect to the main axes of the lattice. For the rectangular lattice this term is $\left(1-\sin^2(2\psi)/2\right)u_{xxx}$. Thus, upon all reflections the soliton front should preserve its orientation with respect to the lattice or change it by $\pi/2$; otherwise the reflected pulse will not remain a stationary soliton due to the change of dispersion.

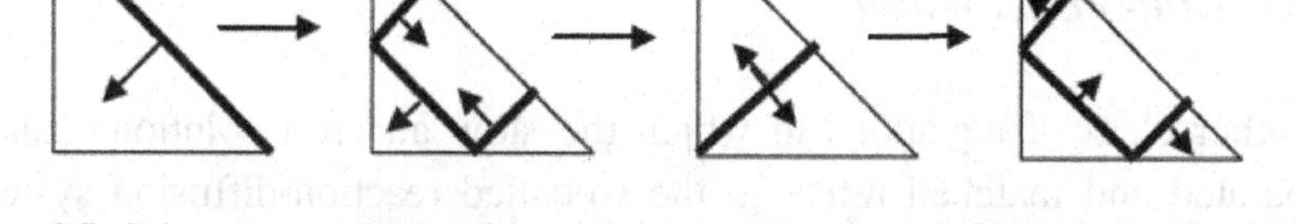

Figure 8.7. Subsequent positions of fronts in two possible types of impulse modes (upper and lower) in an isosceles, right-angle triangle. T is period of oscillations.

Due to this anisotropy, out of the four aforementioned configurations, only the first two meet this condition. Figure 8.7 shows plane fronts in an isosceles, right-angle triangle, where two different resonant "impulse modes" are possible. Some examples of solitons generated in a two-dimensional resonator [15] are shown in Fig. 8.8.

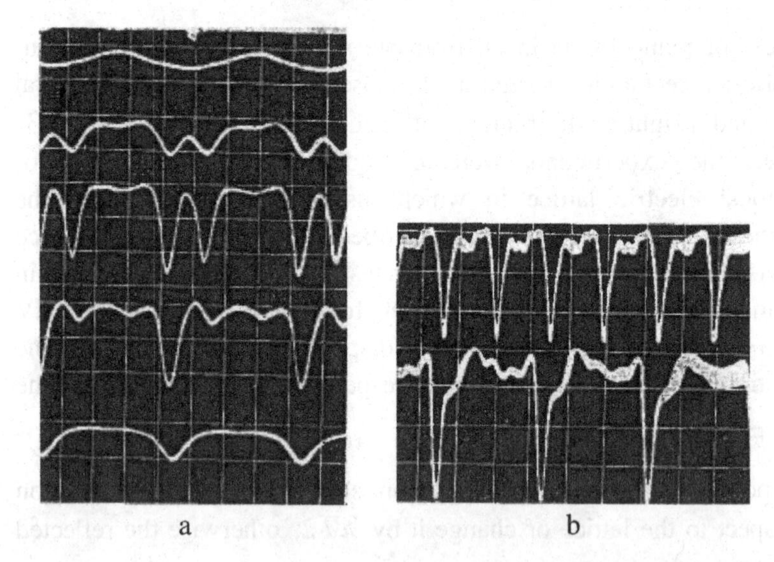

Figure 8.8. (a) Oscillations in a corner of a triangle resonator at different "detunings" between the resonance frequency of a mode and the pump frequency. (b) Period doubling in the same resonator excited at a second harmonic of the main resonance frequency after a small change of excitation frequency.

These results can be extended to three-dimensional problems such as the description of solitons-like excitations in solid crystals.

8.4. Autowaves in Reaction–Diffusion Systems

8.4.1. *KPP–Fisher model*

Another class of equations in which the slow and fast solutions can be separated and matched refers to the so-called reaction-diffusion systems supporting autowaves. This class covers a variety of phenomena: combustion, lasers, chemical and biological systems, etc.; see, for example, the books [18] and [2]. Active studies of reaction-diffusion equations began from the work by Kolmogorov *et al.* [10], who introduced the so-called Kolmogorov–Petrovskii–Piskunov (KPP) equation:

$$u_t = Du_{xx} + F(u), \tag{8.32}$$

where D is a positive diffusion coefficient, and the function $F(u)$ has at least two zero points (equilibria). First we consider the case $F = u\ (1-u)$ which was studied by Fisher [3]:

$$u_t = Du_{xx} + u(1-u). \tag{8.33}$$

In this and other reaction-diffusion equations the processes may be fast or slow separately in time and space, so that we do not introduce universal slow time and coordinate but consider several characteristic cases.

Equation (8.33) has a family of stationary solutions, $u = U$ $(\zeta = x - Vt)$, satisfying the ODE:

$$DU_{\zeta\zeta} + VU_{\zeta} + U(1-U) = 0. \tag{8.34}$$

There exist two equilibrium points, $U_1 = 0$ and $U_2 = 1$, and a transitional front between these two points. Here we are interested in the case $D \sim \mu << 1$. In this case the motions can be divided into "slow" and "fast" in space. The corresponding phase portrait of Eq. (8.34) is shown in Fig. 8.9. The separatrix corresponds to the physically most interesting slow motion, for which the term with D can be neglected (it can be done if $D << VL = V^2\tau$, where L and τ are the spatial and temporal scales of the wave, respectively). The trajectories starting far apart from the separatrix are almost vertical (spatially "fast" motions); they approach the separatrix at a scale of the order of D/V and then remain close to the separatrix.

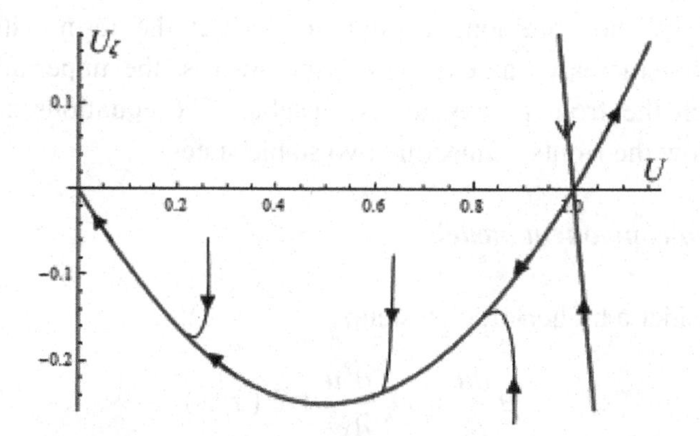

Figure 8.9. Phase plane of (8.34) at small D. Arrows show the direction along ζ. The separatrix going from $U = 1$ to $U = 0$ corresponds to the front shown in Fig. 8.10 below. A few fast motions merging with the separatrix are also shown.

The front corresponding to the separatrix is shown in Fig. 8.10.

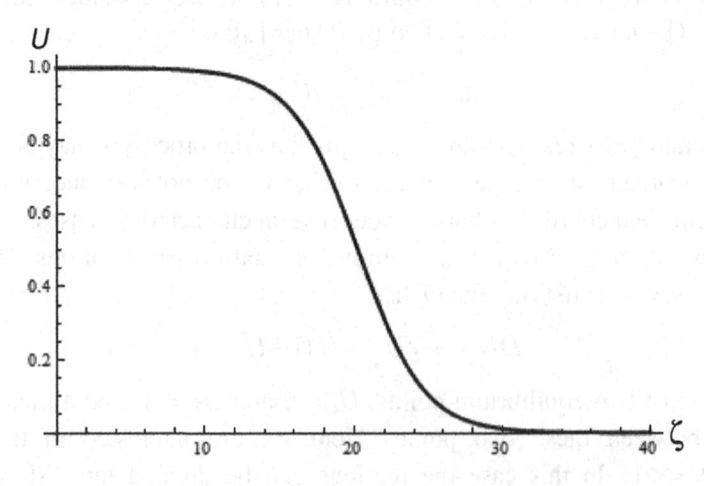

Figure 8.10. Profile of the transient front in Eq. (8.34) at $V = 2.2$ and $D < 0.01$.

Linearization of (8.34) near the equilibrium points easily shows that the aperiodic front exists if $V > 2$. It can also be shown that this front is stable within the framework of the stationary equation (8.34). However, in the framework of Eq. (8.33), for non-steady processes one should consider small deviations from the constant background, $U = 0$, far ahead of the steady front. In this area linear perturbations increase in time; in particular, if they are long enough to neglect the term with D, the perturbation increases as $\exp(t)$. In other words, the unperturbed state into which the front propagates is unstable. The equations considered below allow the fronts connecting two stable states.

8.4.2. *Two-component models*

Consider a higher-order system:

$$\mu\frac{\partial u}{\partial t} = D_u\frac{\partial^2 u}{\partial x^2} + \Psi(u, w),$$
$$\frac{\partial w}{\partial t} = D_w\frac{\partial^2 w}{\partial x^2} + \Phi(u, w). \tag{8.35}$$

Here the parameter μ is again small. Such equations are widely used in biophysics and ecology; in particular, they are closely related to the so-called FitzHugh–Nagumo equations describing excitation waves in nerve fibers (e.g., [8]). Following [13], we suppose that $\Phi = -\gamma u - w$ and $\Psi = w - f(u)$, where $\gamma =$ const and $f(u)$ has an N– shape as shown in Fig. 8.11. An important difference from (8.34) is that Ψ can have three zeros rather than two, which allows the existence of a fully stable front.

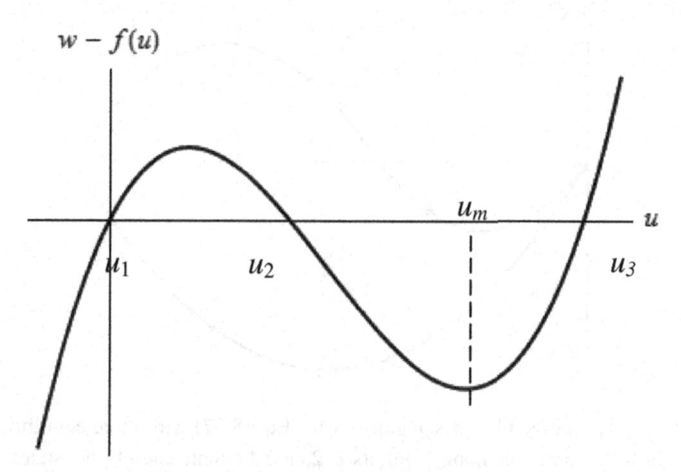

Figure 8.11. The form of the nonlinear function $\Psi = w - f(u)$ at a fixed w. $u_{1,2,3}$ are the equilibrium values of u at the given w.

At $\mu << 1$ we can now distinguish processes which are fast and slow in time (not necessarily in space). For the fast part of the motion $\partial w / \partial t << \partial u / \partial t$, and w can be assumed constant in time, to have

$$\frac{\partial u}{\partial s} = D_u \frac{\partial^2 u}{\partial x^2} + \Psi(u, w), \quad w = w_0 = const, \ s = t / \mu, \qquad (8.36)$$

where s is the "fast" time.

As before, there exists a stationary solution U depending on $\zeta' = x - Vs$ and satisfying the equation

$$D_u \frac{d^2 U}{d\zeta'^2} + V \frac{dU}{d\zeta'} + w_0 - f(U) = 0. \qquad (8.37)$$

We are again interested in separatrices making a transition between the equilibrium points in which $f = w_0$. They are shown in Fig. 8.12; here, as above, the diffusion coefficient D_u can be small so that the motions fast in time can be slow in space. In the Fig. 8.12, point 2 is unstable as was discussed above for the zero equilibrium point in Fig. 8.10. Points 1 and 3, as well as the front corresponding to the separatrix connecting these points, are stable. At a given w_0 the front has a unique velocity $V(w_0)$.

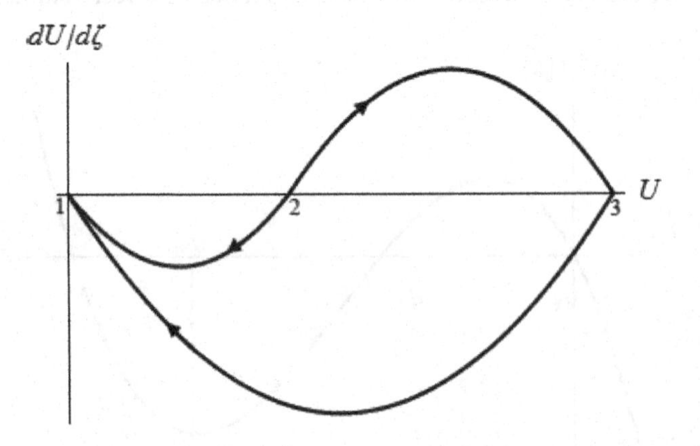

Figure 8.12. Qualitative plot of separatrices for Eq. (8.37) with three equilibrium points. Arrows show the direction along ζ. Points 1, 2, and 3 denote equilibrium states.

Note that there exists a similar solution propagating at the velocity $-V$, in the opposite direction. It is also important that V can change its sign depending on w_0 and for some $w_0 = w_{cr}$, $V = 0$. Integrating (8.37) once at $V = 0$, it is easy to see that the corresponding immovable solution exists at $w_0 = w_{cr}$ defined by the condition

$$\int_{(1)}^{(3)} \Psi(u, w_{cr})du = 0, \text{ or } \int_{U_1}^{U_3} f(u, w_{cr})du = U_3 w_{cr}. \tag{8.38}$$

Here the points 1 and 3 correspond to the equilibrium quantities U_1 and U_3, respectively.

In [13] the dependence $V(w_0)$ was calculated using a piecewise-linear approximation of the function $f(u)$. If the front velocity V is close to zero, the fast solution approximation defined by Eq. (8.36) is inapplicable, and the full system (8.35) should be analyzed.

In the other limit, that of the motion slow in time, the term with μ can be neglected, and system (8.35) acquires the form

$$D_u \frac{\partial^2 u}{\partial x^2} + \Psi(u,w),$$
$$\frac{\partial w}{\partial t} = D_w \frac{\partial^2 w}{\partial x^2} + \Phi(u,w). \tag{8.39}$$

In this system we suppose that the diffusion coefficient D_u is small, so that spatial variations of u can be "fast", much steeper than those of w.

Consider now a combined process: evolution of an initial pulse. Suppose that the initial perturbation is localized but smooth so that its spatial scale is large as compared with the diffusion scales, $L_{u,w} = \sqrt{\tau_s \cdot D_{u,w}}$, where τ_s is the characteristic time of the process, and that $w(x,0) > w_{cr}$. For simplicity we suppose that $w(x, 0) = w_{eq} =$ const(x) and w_{eq} corresponds to its equilibrium value. The subsequent motion consists of the following phases (see Fig. 8.13).

1. Under the above assumptions, at the first stage the diffusion can be neglected, and at the constant w, the part of the wave exceeding U_3 reaches U_3, while the part with $u < U_3$ decreases to U_1, both in the fast time interval. As a result, the pulse becomes flat-top with sharp edges (Fig. 8.13a) corresponding to stationary fronts described by (8.37) and propagating in opposite directions, so that the pulse expands. If the initial pulse is symmetric, it remains symmetric thereafter, so that it is sufficient to describe the evolution of, e.g., its right half.
2. The front moves with a constant velocity $V_1 = V(w_{eq})$. The field behind the front now varies independently in each point due to the slow motion (Fig. 8.13b) which involves variation of w and satisfies the Eqs (8.39) with D_u neglected:

$$\frac{\partial w}{\partial t} = \Phi(u,w), \quad \Psi(u,w) = 0. \tag{8.40}$$

After substituting the above expressions $\Phi = -\gamma u - w, \quad \Psi = w - f(u)$, the integral of (8.40) has the form

$$w = f(u),\ t = \int_{u}^{u_3} \frac{f'(u)du}{\gamma u + f(u)} = t_u(u, u_3). \tag{8.41}$$

In this solution u and w monotonously decrease up to the values u_m and $w_m = f(u_m)$ corresponding to the minimum point on the curve shown in Fig. 8.11.

3. Subsequently the variable u jumps down (at a constant $w = w_m$), forming a new fast transition of an opposite sign moving in the same direction as the front but with a different velocity, $V_2 = .V(w_m)$. These two fast transitions form an excitation pulse. This process is shown in Fig. 8.13c. If $V_2 > V_1$, the pulse becomes shorter, but w eventually increases, and at $t \to \infty$, asymptotically $V_2 \to V_1$, so that the pulse length stabilizes at a finite value. If $V_2 < V_1$, which is possible if the function $f(u)$ is strongly asymmetric, after the fast formation of the rear jump, the velocity of the frontal and rear jumps turn out to be equal from the very beginning, and the pulse remains stationary.

4. Behind the pulse the system slowly returns to the equilibrium state (in biology it is called the refractory phase), Fig. 8.13d.

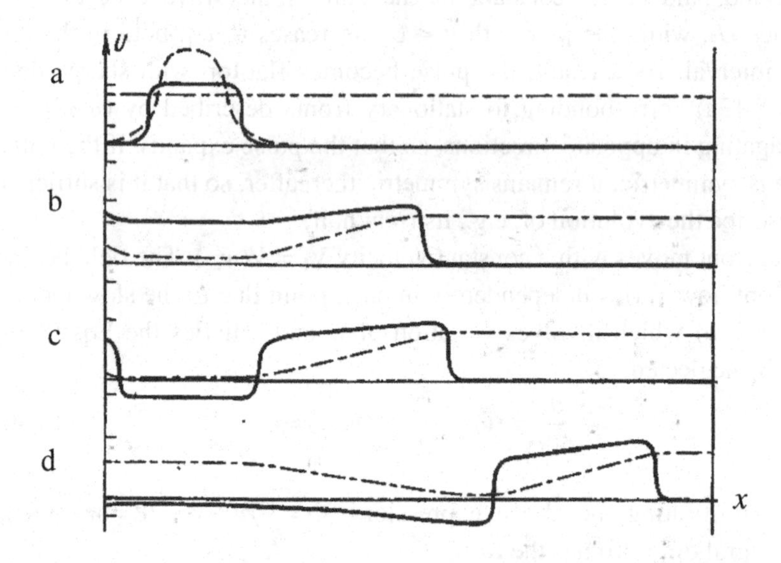

Figure 8.13. Formation of excitation pulse at $w > w_{cr}$. Dashed line: – initial u -pulse. Solid line: – u, dash-dotted line: – w. The full picture is symmetric with respect to the middle of the initial perturbation (dashed line on panel a).

Hence, the process ends with the formation of two diverging, stationary pulses of excitation.

The scenario described above occurs if $w_{eq} > w_{cr}$. If $w_{eq} < w_{cr}$, then again a rectangular pulse forms at stage 1, but the pulse edges move towards the center, so that the perturbation shrinks and disappears in a finite time. This qualitatively corresponds to the "mutual killing" of two oppositely moving fronts of flame or excitation fronts in a nerve fiber.

This theory can be readily extended to axially symmetric two-dimensional systems by using operators $r^{-1}(r\partial/\partial r)$ for the diffusion terms. As a result, a pulse similar to that shown in Fig. 8.13 but having a circular front can be observed, as well as more complex, spiral structures (e.g., [7]). A review of relevant publications was made by Tyson and Keener [17].

8.5. Concluding Remarks

In this chapter we discussed examples of the broad class of dissipative waves and autowaves, in order to illustrate their specific features. In some cases, such as the slow growth of a soliton, extension of the previous schemes to active systems is straightforward; in other cases, such as an important class of reaction-diffusion PDEs, a consistent asymptotic procedure is not yet sufficiently developed.[a] Here we presented the description of such processes based on separation of fast and slow processes in space and time and their matching, including formation of steady-state or slowly varying excitation fronts and impulses. It can be stated, however, that the obtained solutions correspond to the main approximation of the perturbation theory. This important branch of asymptotic perturbation theory is progressing.

---

[a] There exist some formal perturbation and iteration schemes developed for the reaction-diffusion equations (e.g., [9]), where a stationary variant of such equations is considered), but no specific dynamical processes were studied.

References

1. Courant, R. and Friedrichs, K. O. (1976). *Supersonic Flow and Shock Waves.* Springer, New York and Berlin.
2. Debnath, L. (2005). *Nonlinear Partial Differential Equations for Scientists and Engineers*. 2nd Ed. Birkhauser, Boston.
3. Fisher, R. A. (1936). The wave of advance of advantageous genes, *Annals of Eugenics*, v. 7, pp. 335–369.
4. Gaponov, A. V., Ostrovsky, L. A., and Freidman, G. I. (1967). Shock Electromagnetic Waves. *Radiophysics and Quantum Electronics,* v. 10, pp. 772–793.
5. Gash, A., Berning, T., and Jager, D. (1986). Generation and parametric amplification of solitons in a nonlinear resonator with a Korteweg–de Vries medium, *Physics Review A*, v. 34, pp. 4528–4531.
6. Gorshkov, K, A., Ostrovsky, L. A., and Papko, V. V. (1973). Parametric amplification and generation of pulses in nonlinear distributed systems. *Radiophysics and Quantum Electronics*, v. 16, pp. 919–926.
7. See Reaction-diffusion system, *Wikipedia*, (accessed 15/12/13) http://en.wikipedia.org/wiki/Reaction-diffusion_system.
8. Izhikevich, E. M. (2007). *Dynamical Systems in Neuroscience: The Geometry of Excitability and Bursting.* MIT Press, Cambridge, MA.
9. Kalachev, L. V. and Seidman, T. I. (2003). Singular perturbation analysis of a diffusion/reaction system whose solution exhibits a corner-type behavior in the interior of the domain diffusion/reaction system with a fast reaction. *Journal of Mathematical Analysis and Applications*, v. 288, pp. 722–743.
10. Kolmogorov, A. N., Petrovskii, I. G., and Piskunov, N. S. (1937). Study of a diffusion equation matched with the increase the quantity of matter and application to one biological problem, *Moscow University Mathematics Bulletin*, v. A1, pp. 1–26. [translated into English and reprinted in Pelce, P. (1988). *Dynamics of Curved Fronts*. Academic Press, San Diego.]
11. Kolomensky, A. A. and Lebedev, A. N. (1966). *Theory of Cyclic Accelerators.* North-Holland, Amsterdam.
12. Naugolnykh, K. A. and Ostrovsky, L. A. (1998). *Nonlinear Wave Processes in Acoustics*. Cambridge University Press, Cambridge.
13. Ostrovskii, L. A. and Yakhno, V. G. (1972). The formation of pulses in an excitable medium, *Biophysics,* v. 20, 498–503.
14. Ostrovsky, L. A. and Potapov, A. I. (1999). *Modulated waves. Theory and Applications*. Johns Hopkins University Press, Baltimore.
15. Ostrovsky, L. A., Papko, V. V., and Stepanyants, Y. A. (1980). Solitons and non-linear resonance in two-dimensional lattices, *Journal of Experimental and Theoretical Physics*, v. 78, pp. 831–841.

16. Smith R. E., Smettem K. R. J., Broadbridge P., and Woolhiser D. A. (2002). *Infiltration Theory for Hydrologic Applications.* American Geophysical Union, Washington.
17. Tyson, J. J. and Keener, J. P. (1988). Singular perturbation theory of traveling waves in excitable media, *Physica D*, v. 32, pp. 327–361.
18. Vasilyev, V. A., Romanovsky, Y. M., Chernavsky, D. S., and Yakhno, V. G., (1987). *Autowave Processes in Kinetic Systems.* VEB Deutcher Verlag der Wissenshaft, Berlin.
19. Whitham, G. B. (1974). *Linear and Nonlinear Waves.* Wiley, New York.

Index

Printed in the United States
By Bookmasters

Printed in the United States
By Bookmasters